3 1331 00795 1538

D0852656

About Island Press

Since 1984, the nonprofit organization Island Press has been stimulating, shaping, and communicating ideas that are essential for solving environmental problems worldwide. With more than 800 titles in print and some 40 new releases each year, we are the nation's leading publisher on environmental issues. We identify innovative thinkers and emerging trends in the environmental field. We work with world-renowned experts and authors to develop cross-disciplinary solutions to environmental challenges.

Island Press designs and executes educational campaigns in conjunction with our authors to communicate their critical messages in print, in person, and online using the latest technologies, innovative programs, and the media. Our goal is to reach targeted audiences—scientists, policymakers, environmental advocates, urban planners, the media, and concerned citizens—with information that can be used to create the framework for long-term ecological health and human well-being.

Island Press gratefully acknowledges major support of our work by The Agua Fund, The Andrew W. Mellon Foundation, Betsy & Jesse Fink Foundation, The Bobolink Foundation, The Curtis and Edith Munson Foundation, Forrest C. and Frances H. Lattner Foundation, G.O. Forward Fund of the Saint Paul Foundation, Gordon and Betty Moore Foundation, The Kresge Foundation, The Margaret A. Cargill Foundation, New Mexico Water Initiative, a project of Hanuman Foundation, The Overbrook Foundation, The S.D. Bechtel, Jr. Foundation, The Summit Charitable Foundation, Inc., V. Kann Rasmussen Foundation, The Wallace Alexander Gerbode Foundation, and other generous supporters.

Unnatural Selection

Library of Congress Control Number: 2014939900

Printed on recycled, acid-free paper

Manufactured in the United States of America
10 9 8 7 6 5 4 3 2 1

Keywords: rapid evolution, antibiotic resistance, vaccines, pesticide resistance, Roundup, cancer treatment, bedbugs, toxics, epigenetics

Contents

Acknowledgments

Most of these chapters touch upon toxic chemicals in one way or another; all of them refer to evolution—a topic that I have just begun to explore. For providing me with the opportunity to think about toxic chemicals and evolution in our lifetime, or rapid evolution, I would first like to thank Island Press and my editor Emily Davis. Emily has helped me rein in my tendency to head off along some marginally related tangent, or cram each paragraph with all sorts of fascinating but not exactly relevant scientific tidbits.

Throughout this process I have relied upon the contributions of many scientists, using their published research and, beyond that, calling upon them to be interviewed, or to review chapters, portions of chapters, or answer questions from, in most cases, a complete stranger. Many offered further suggestions, corrections, and encouragement though an exchange of e-mails. Despite their best efforts I am sure many omissions, mistakes, or inaccuracies remain, for which I am fully responsible. The list of those scientists who kindly contributed to this book includes but is not limited to: Josh Akey, Claude Boyd, Steven Brady, Adria Eiskus, Suzanne Epstein, Amir Fathi, Marco Gerlinger, Greg Jaffe, Norman Johnson, Paul Klerks, Katia Koelle, Ben Letcher, Emmanuel Milot, Mike Owen, Colin Parrish, Rick Pilsner, Andrew those large changes are subtle ones: a slightly longer finger bone in a bat; a minor change in the pattern on a butterfly wing; a deeper purple in a flower's petal. This is evolution writ small.

If we could peer into life's inner workings, whether bacteria, bug, fish, or frog, we would see that the very code of life is more fluid than once imagined. Most often, changes in DNA or its expression are unnoticeable or unimportant, but every once in a while, a doozy of a mutation pops up: a lethal strain of bacteria gains the ability to chew up and spit out the last effective antibiotic; or a malaria-carrying mosquito brushes aside DDT, unaffected; or the daughter of prehistoric farmers finds she is able to tolerate milk into adulthood. While such variations may well be the spice of life, evolution requires favorable conditions for the selection of these new flavors. A mutation that detoxifies DDT is only helpful when mosquitos are under pressure from insecticide; digesting milk only helps where milk is available and when drinking it contributes to a brighter future for the whole family. And a trait's surviving under pressure is of little use to the larger population unless it is a trait that can pass from one generation to the next. When beneficial heritable characteristics, whether through mutation or other means, arise in the right place, the right time, and in the right species, they may sweep through populations.

This is evolution in the fast lane, and species with explosive population growth—the bugs, bacteria, and weeds (or in the case of cancer, cells)—have the advantage over those of us who reproduce more slowly. When mosquitos, bedbugs, and houseflies evolved resistance to DDT in the virtual blink of an eye, other species such as eagles, peregrines, and pelicans faced extirpation. We won't see the evolution of tusk-free African elephants in heavily hunted populations, or contaminant-resistant polar bears (as top predators, polar bears are among the most contaminated animals on earth) in "contemporary" time, but we are certain to encounter plenty of chemically resistant pests and pathogens.

This book begins with our relationship to those species living life in the fast lane. Some of the more infamous cases of rapid evolution are introduced in the first section, "Unnatural Selection in a Natural World." Organisms like staph bacteria, agricultural weeds, and even bedbugs are evolving in direct response to our actions. It is a completely natural phenomenon, but it's occurring in a human-tainted unnatural setting. Remove the human element and you remove a good portion of the selection pressure. Each chapter of this first section reveals a different facet of our experience with rapid evolution, from discovery to prevention to resurgence of species once controlled by our chemicals. By better understanding the processes at work, we can do a better job preventing or fixing the ensuing problems. Rather than risk heading off into a near future filled with "superbugs," we can change how we interact with pests and pathogens, reduce the pressure, and still maintain some degree of control.

Though we tend to be most concerned with how evolution directly impacts *us*, it is becoming increasingly clear that our evolutionary influence extends well beyond bugs and bacteria to wildlife. In rivers, ponds, and lakes toxic enough to kill—whether contaminated with pesticides, toxic metals, or PCBs—the flash of a fin, clutch of eggs, or frog song is signaling that rapid evolution isn't just for the spineless. Species with a backbone are also evolving in response to chemically altered environments. When I first proposed this book, reviewers expressed concern that writing about evolution in response to pollution would provide industries fighting billion-dollar cleanups and pollutant controls with an excuse—if contaminated fish are fat and happy, why bother cleaning up?

By the end of this book I hope the answer will be clear. Those surviving fish will have subtle genetic differences from their pre-industrial-age ancestors. Yet how those differences—the result of our chemical influence—will play out in the long run is anyone's guess. What happens when tolerant fish or frogs loaded with chemical contaminants pass their toxic burden on to more sensitive species like the hawks and minks that feed upon them? Or when survival comes with a price tag? Perhaps a toxic-tolerant population is more sensitive to temperature swings, starvation, or predation. Some of those species surviving today may be gone tomorrow. We are laying the groundwork for a game of *Survivor: Planet Earth*, and the outcome may not be to our liking. With a century of industrial-age pollutants behind us, and billions of pounds of toxicants released into air, water, and land each year, the stories of fish, frog, and salamander told in the second section of this book, "Natural Selection in an Unnatural World," are unfolding as you read. As long as industry remains wedded to the false dichotomy of profit over protection, and we keep choosing cheap over sustainable goods, pollutants will continue to settle near and far across the globe, changing life.

If the rapidity of evolution in response to drugs, pesticides, and pollutants isn't enough to make us think twice about our chemical dependence, here is another reason: epigenetics, the heritable changes in gene *expression*, without any changes in the DNA sequence itself. Evidence is piling up that some environmental stressors leave their mark on plants and animals, including humans, for generations to come by altering how and when genes are turned on and off. The last section, "Beyond Selection," offers a speculative yet disturbing scenario. Epigenetic change may provide an incredibly rapid source of variation in a single generation. And the stressors shown to cause epigenetic changes range from temperature to nutrition to toxic chemicals. If variation induced by stress, including chemical exposures, can pass from one generation to the next—could these changes influence the course of evolution? And if so, how might exposed species, including ourselves, fare?

We cannot turn back the clock, nor would most of us want to return to pre-industrial, pre-antibiotic days. But we can learn to live in balance—to manage pests without creating insects invulnerable to our safest and most effective insecticides; to protect individuals from disease without inviting epidemics; to benefit from technology and modern chemistry without threatening the health of future generations. The first step is understanding how our choices impact life's evolutionary course. And so we begin close to home, with an impending public health disaster: antibiotic resistance.

CHAPTER 1

Discovery: Antibiotics and the Rise of the Superbug

"I see resistant staph all the time," says nurse practitioner Maggie G. Her enormous blue eyes convey both the compassion and the weariness of someone who has seen it all. Over the course of 25 years, the Western Massachusetts nurse has treated farmers, hill-town hippies, and teens seeking treatment for STDs and fevers, as well as men, women, and children who walk for miles and wait patiently with festering wounds and suppurating tumors in the Sierra Leone clinic that she visits once a year. One constant throughout all of Maggie's experiences is methicillin-resistant staph, or MRSA. Back in the late eighties, when Maggie was just finishing nursing school, MRSA was rare. But over the years she has witnessed the rise of this drug-resistant bug, tending to countless cases—one of the most memorable involved a young camp counselor whose infected toe turned into a life-threatening hole in her heart. When we spoke, Maggie was working with recovering addicts at a psych hospital. MRSA spreads so easily in needle-using addict populations through needle sharing or festering open wounds that Maggie says addicts are often treated "presumptively"—meaning the staff doesn't always test but assumes drug resistance. It's a reasonable assumption. In some places, nearly 50 percent of the needle-using population may be positive for community-acquired MRSA.[1]

First recognized as a "healthcare-associated infection" limited to patients and caretakers, MRSA made its way out of the hospital into the community a decade or so ago. The bacteria can spread from mother to daughter, throughout a high school locker room by way of an infected towel, from pet to owner, and between hospital patients on the hands of a caregiver. It is a parent's worst nightmare: a small bite or scrape turns into an angry red trail streaking up a child's leg, and one antibiotic after another fails. A once easily treatable infection is now potentially fatal. Of the roughly 75,000 Americans who become infected with MRSA each year, an estimated 9,700 will die.[2]

We live in dangerous times. Infectious diseases are rapidly evolving beyond our medicinal reach, returning us to the pre-antibiotic age. In just over a century, we have rendered impotent some of our most precious therapies, and there is plenty of blame to go around. Whether it be doctors pacifying pushy, anxious parents; the agricultural industry preventively treating livestock, or worse, simply encouraging livestock growth; or hospitals fending off recalcitrant infection—we have all contributed to the rise of the superbug. Each year nearly 37 million pounds of antibiotics are used in the United States. Some 7 million pounds go down the throats of our kids, up the arms of hospital patients, and into infected addicts; a few hundred thousand pounds are consumed by our pets; and the rest is used by the ag industry.[3] And though MRSA is the poster-bug for resistance, it has plenty of company. A once-curable pneumonia recently killed seven patients at a well-regarded national hospital.[4] Tuberculosis that is completely drug resistant has surfaced in India, Italy, and Iran. In Japan, a strain of gonorrhea has shaken free from all antibiotics. That a fully antibiotic-resistant STD may once again rage throughout the world ought to strike fear into all of us, even those who consider ourselves beyond its reach—if not simply because "you just never know," then because some bacteria can easily swap resistance genes. And that means that resistance in a venereal disease may one day transfer to a bug that causes pneumonia or a skin infection. Bacteria may be among the most primitive life forms on earth, but they have proven to be among the most formidable opponents.

The story of antibiotic resistance is one of great advances and impending loss. It begins a little over a century ago with two of the most important discoveries in modern medicine: that disease can be caused by bacteria, and that bacteria can be killed selectively. Yet almost as soon as antibiotics hit the market, one after another began to fail. Antibiotic resistance touches us all, so it is a good place to begin an exploration of evolution in our time.

Discovery

Before we can cure (or better, prevent) disease, we must recognize the cause. The French chemist Louis Pasteur was just a boy in 1831 when cholera killed nearly 20,000 souls in Paris, roughly 250 miles (400 kilometers) from his hometown. These were the days when epidemics raged—infecting and killing until new victims were too few in number to sustain the spread of disease. Cholera killed millions worldwide. Bubonic plague flared up every few hundred years or so, at one point taking over a third of Europe's population. There was plenty of disease to go around, particularly in dense and well-traveled populations. When Pasteur entered the world, physicians and scientists attributed the cause of infectious disease in large part to "miasma"—poisonous vapors in the air. Disease, like the winds, weather, and the stench of a fetid river, seemed to travel, hovering here and there for days, months, or years before moving on. A few practitioners insisted that disease spread through contact as a tangible entity rather than some amorphous gas—but without proof, the miasma theory ruled. Louis Pasteur's research eventually focused on revealing the invisible causes of disease, moving

who envisioned a day when medicine wouldn't just prevent but *cure*, leaving the host alive and well. Just one year before Koch's death, the first antibiotic drug was introduced to the world.

Target Practice

Antibiotic (or, more accurately in this case, anti*bacterial*) use is chemical warfare waged on a microscopic field. The trick is to destroy the pathogen, yet leave our own cells unharmed. But there is a catch. Having shared a common ancestry for over a billion years, our cells have much in common with bacteria, which makes identifying a specific target akin to playing "Spot the Difference." Even in these days of genomics and advanced analytical chemistry, it is a difficult game. Singling out bacteria at the turn of the last century would have been like playing the game blindfolded.

And so scientists exploited the very few differences they *could* see. By the late 1800s, industrial chemical dyes had begun to make the invisible visible, tagging biological structures with red, blue, and purple. For the first time ever, a physician could both differentiate animal cells from bacterial cells *and* distinguish one class of pathogen from another. If chemical dyes clung to one cell type while ignoring another, could these chemicals also be used to kill pathogens while leaving host cells unharmed? Was there a way "to aim chemically"?[7]

Turn-of-the-century German physician Paul Ehrlich was the first to investigate this central question. Connecting the dots between chemistry, bacteriology, and medicine, Ehrlich assembled a team and took aim at one of the more infamous diseases of the day, syphilis. After running hundreds of chemicals synthesized by the German dye industry through their paces, the scientists discovered that compound number 606 was the winner, curing infected rabbits with a single dose.[8] It was 1909, and within a year number 606, renamed Salvarsan, made its way to the clinic. Syphilis, a chronic and potentially fatal disease for the ages, had become curable.[9] Salvarsan singled out a pathogenic bacterium and, when it was administered properly, caused relatively little collateral damage.[10] The antibiotic age had just begun: synthetic products were poised to make their way into medicine cabinets, hospitals, and our bodies. The uneasy relationship between human and bacterial pathogen, shaped by millions of years of coevolution, was about to change. But it would take two world wars before humans finally gained the advantage.

As the twentieth century dawned, Pasteur's, Koch's, and even Ehrlich's discoveries notwithstanding, eating, drinking, or getting a simple puncture wound could still send one to their grave. Syphilis was but one disease, and hygiene could only go so far in disease prevention. Infections we rarely think about today—cholera, typhus, strep, and staph—continued to run their course, killing and maiming. For pathogens, World War I, like so many other wars, was a war of opportunity. As bullets and bombs shredded skin and tore limbs from bodies, infectious bacteria thrived. Aspiring physicians had few options but to amputate infected limbs and watch helplessly as young men died. If they didn't die from infected wounds, there were plenty of other diseases, like cholera and typhus, waiting in the wings. Gerhard Domagk, a volunteer and medic in the German army, had ample opportunity to observe the quick work that bacteria made of men. Years after the war, inspired by Ehrlich's vision of a "magic bullet" cure, Domagk turned his attention to finding the magic in industrial dyes. The target was *Streptococcus*, a common cause of skin infections that could quickly take a turn for the worse. One red dye proved particularly effective at curing strep-infected mice, yet any fanfare would have to wait for human testing. But before those tests could be completed, an odd twist of fate intervened. Domagk's six-year-old daughter, Hildegard, fell ill with a life-threatening infection. She had punctured her hand with a sewing needle. Hospitalized with fever and infection progressing up her arm, she faced the standard treatment—amputation with no guarantee of survival. Desperate, Domagk treated Hildegard with the dye. Days later she recovered. It soon became apparent that the dye targeted not just strep but a number of other infections as well. Within a few years, the dye, packaged and sold as Prontosil, became the first commercially available sulfa drug. Its derivatives remain in use today. The discovery offered a cure for illnesses from child-bed fever to pneumonia, skin infections, and gonorrhea. It was 1935, and for the first time in human history a whole range of once-fatal infections could be cured.[11] Less than five years later, Domagk took home a Nobel Prize.

Two chemicals discovered nearly two decades apart, both products of a chemical industry delirious with newfound ability to synthesize novel chemicals and scientists willing to test one after another, offered a world of change. Yet evolution had already produced a far more effective antibacterial chemical, as Scottish physician Alexander Fleming would discover.

Penicillin

Like Domagk, Fleming returned home from World War I bent on disease prevention, only to discover by sheer accident one of the most valuable antibiotics of the century. His is the now-classic story of accidental discovery: a summer vacation trip, stacks of petri dishes dotted with colonies of staph bacteria left in the sink, an observation, and the historic follow-up. Cleaning the lab after returning from vacation, Fleming noticed that an invisible conflict was playing out on plates that been left to molder. Where spots of one particular mold contaminated the plates, bacterial colonies failed to grow. Fleming's genius was to ask why this particular mold (subsequently identified as a *Penicillium*) cleared the surrounding bacteria. Follow-up studies showed that it produced soluble chemicals that killed not only staph but an assortment of other bacteria.

What Fleming couldn't have known was that penicillin hit bacterial cells where it really hurt—the cell wall. Like a chicken-wire frame for a papier-mâché sculpture, the wall protects bacterial cells from bursting under their own internal pressure. If the wall is compromised, bacteria pop like overfilled balloons. Bacteria like staph and strep with thick cell walls are most sensitive, while others with thinner cell walls, like *Salmonella* and *coliform*, are less sensitive. Our animal cells lack cell walls and so avoid damage. Penicillin is a tribute to the ingenuity of nature. When bacterial cells grow and divide, the wall is broken down and rebuilt. Penicillin prevents that molecular frame from linking back together.

Penicillin's discovery would have been a watershed moment for antibiotic development. But available technology was insufficient for isolating or producing quantities of active chemical, preventing Fleming from putting it through its paces. He published his findings in 1929, but the work slipped quietly into the literature.[12] Nearly a decade later, it piqued the interest of a trio of scientists: Howard Florey, Ernst Chain, and Norman Heatly. Finally able to produce enough of the "mold juice" to test, the group treated mice infected with a lethal dose of strep. The results were impressive. Untreated mice died. Those given penicillin lived.[13] Human testing followed. In 1941, Police Constable Albert Alexander entered a hospital in Oxford, England, after a rose prick blossomed into a raging, life-threatening infection. Penicillin turned the tide, for a while. Unfortunately for Alexander, no one knew how much of the new drug was needed to beat back an advanced infection. With only enough to treat the constable for several days, despite extracting it back from his urine, Alexander succumbed. Yet this was enough to confirm penicillin's efficacy in humans. Fleming's all-but-forgotten discovery soon took center stage in the war against bacteria. Those early trials, combined with another world war, the desperation of battlefield medics, and the increasing failure of sulfa drugs (the first antibacterials to fall to resistance), helped turn an astute observation into one of the greatest discoveries of the twentieth century. But the triumph was short-lived.

Resistance!

As the Second World War came to a close, the great technology transfer began. From nuclear power to plastics, pesticides, and antibiotics, the era of "Better Living through Chemistry" had arrived. Penicillin was ripe for exploitation. Just as everything today, from mattresses to shopping-cart handles, is impregnated with antimicrobials, industry envisioned toothpaste, chewing gum, lozenges, face creams, and even vaginal *crèmes* infused with penicillin. In his 1945 Nobel acceptance speech, Fleming warned that penicillin's overuse and under-treatment of disease could result in resistant bacteria.[14] But it was already too late. Our first lesson in moderation had come and gone, as resistant strains of staph, strep, and pneumonia cropped up during the war. Penicillin had imposed a powerful selection pressure. Pathogens that could not evolve would die. But in hospital wards both here and in Europe, penicillin-resistant staph began making the rounds, along with sulfa-drug-resistant bacteria. One could almost watch resistance evolve.[15] Attempting to control dangerous strep infections in new recruits, the US Navy treated hundreds of thousands of trainees with prophylactic doses of the drug sulfadiazine. Rheumatic fever, scarlet fever, and respiratory disease incidence dropped almost immediately, but sulfa-resistant strep emerged just three *months* after the initial phase of treatment.[16] Similarly, penicillin was losing ground. As Fleming had feared, one of the greatest factors in the decline of antibiotic effectiveness proved to be overuse.

Then, in 1950, as if to throw fuel on fires of resistance, scientists discovered that antibiotics added to livestock feed accelerated growth, moving animals more quickly from farm to table. Even better, antibiotics helped cut production costs.[17] It was an apparent win-win for the farmers struggling to meet the booming postwar demand for meat and for customers craving an affordable, protein-packed meal. Antibiotics weren't just for the sick and dying anymore—they had become an integral part of "what's for dinner."

The medical world's failure to heed Fleming's warning was a combination of hubris and naïveté about genetics and evolution. Medicine was on a roll: if one antibiotic failed, another would surely take its place. If evolution worked to the bacterium's advantage, human ingenuity would work toward ours. Penicillin was one of the first drugs to be improved. Under penicillin's pressure and through the process of natural selection, bacterial populations acquired a gene for an enzyme capable of snipping apart the drug's key chemical structure—called a beta-lactam ring. In response, drug developers created methicillin. Like having a portcullis to block the castle gate, methicillin contained a molecule that protected the bacterial-busting beta-lactam structure from destruction. This so-called super-penicillin did the trick. That was in 1959. By 1961, reports of methicillin-resistant staph emerged in England and Europe.[18] The invaders had found another way around. By 1968, researchers at Boston City Hospital had isolated 22 methicillin-resistant strains from 18 patients.[19] Most had become infected after admission to the hospital. It was the dawning of the age of hospital-acquired MRSA.

Several modified versions of penicillin followed, as did the discovery of other natural and synthetic drugs including bacitracin, streptomycin, rifampicin, erythromycin, and polymyxin. The majority were discovered during the antibiotic golden years, 1930–1950. Today, they are fast becoming obsolete, and unfortunately, the next best thing isn't around the corner. Many of our current antibiotics, like penicillin, are beta-lactams that inhibit cell-wall synthesis. Some inhibit the production of proteins, and others alter bacterial cell membranes. At prescribed dosages, most target bacteria while leaving our cells intact and relatively healthy. One increasingly recognized downside of antibiotics, though, is their inability to distinguish the pathogenic bacteria from the "good" bacteria, some of which may even help fend off disease. And while most MRSA patients, including those in Maggie's psych clinic, may benefit for now from next-generation drugs, those options are not available to all. "People in Sierra Leone die from infections," says Maggie, weary with resignation. "We're still using first-line antibiotics there—if we even have them."

The majority of medically important microbes now resist one antibiotic or more, and words like "nightmare" and "catastrophic" are increasingly cropping up in the medical literature. It is not just hyperbole.[20] Like Aesop's Hare, whose overconfidence led to a predictable loss against Tortoise, our hubris may very well cost us our health—if not our lives. Certainly our current situation is not for lack of understanding; we know far more about bacteria and evolution than Pasteur, Koch, or Ehrlich could have even dreamed. Yet we continue to play whack-a-mole—simply changing antibiotics as resistance pops up. It is time to reconsider our strategy and pay homage to evolution. Not the dusty old process of evolution that we equate with the descent of man and speciation but the wild, DNA-swapping, mutating ways of bacteria.

Unveiling the Machinations of Evolution

A single *Staphylococcus aureus* cell, like most bacteria, can within days give rise to millions, if not billions of daughter cells. Bacteria reproduce by cloning. The parent cell divides into two daughters that in turn generate their own daughter cells on and on as one cell exponentially yields hundreds, thousands, and then millions of new cells—by any measure, an impressive amount of DNA replication. Not all of it perfect, though. With each new generation comes the potential for mutation. And mutations are a source of variation for evolution. This is true no matter the species, whether bacteria, bedbugs, elephants, or humans. Only some enjoy the advantages of new gene variants in the course of a few months or years while others like us might require centuries. And though most mutations are of little or no benefit, it only takes one alteration in the right place, and *voilà*, an enzyme is no longer a suitable target for chemical attack. When these advantageous traits are selected, evolution happens.

The explosive population growth of bacteria means that a beneficial mutation can infiltrate a population within hours. Contrast that with the thousands of years required for a random yet beneficial mutation to take hold in a human population. When hospitalized patients are treated with antibiotics for weeks or months, there is potential for myriad new (or *de novo*) mutations, which in turn become feedstock for the evolution of resistance. As rare but helpful mutations arise—particularly in bacteria under the influence of antibiotics—resistance isn't futile, it is inevitable. In one case, researchers caught evolution in action as the staph infecting a patient treated with vancomycin acquired 35 sequential mutations, diminishing the antibiotic's efficacy as mutations accrued.[21]

As impressive as rescue by *de novo* mutation may be, bacteria have an even more efficient means for acquiring resistance. For so-called sexless organisms, bacteria are incredibly agile genetically. Japanese researchers were the first to catch a glimpse of the acrobatics. During the Second World War, and in the years that followed, *Shigella dysenteriae* became epidemic in Japan. Even if dysentery didn't kill, it knocked the survivors flat. Sulfa drugs worked at first, but by the early 1950s, *Shigella* had evolved resistance. Then Japanese researchers observed something that should have beat some sense into scientists and physicians around the globe. Some strains of *Shigella* were resistant not only to sulfa drugs but to other, newer, drugs as well.[22] They would be the first reported multi-drug-resistant bacteria. The response by the Western medical world was underwhelming. Evolution of *any* resistance over the course of treatment was believed to be a low-probability event. According to antibiotic pioneer Julian Davies, "The notion of multiple drug resistance was heretical."[23] And that wasn't all.

A scientist working in the United Kingdom had isolated bacteria that were, oddly enough, resisting *novel* antibiotics right off the bat.[24] Resistance, it seemed, had spread from one strain to another. Follow-up studies suggested that bacteria were sharing resistance through contact. The findings, as Davies recounts, challenged the prevailing ideas about the process of evolution. *If* evolution simply proceeded by way of one random mutation at a time, and resistance required selection pressures like an antibiotic, then how could these findings be explained? As with the finding of multi-drug-resistant *Shigella*, the reception was unenthusiastic at best, doubtful at worst.[25] But then, how else could resistance to so many drugs evolve so quickly? And why would bacteria carry resistance to a novel antibiotic? The answer lies in the so-called sex lives of bacteria.

When bacteria reproduce, much like us their genes are handed down from parent to offspring, vertically. Just as we carry our genes on linear double-stranded chromosomes, bacteria, too, carry their genes on a double-stranded chromosome—but the bacterial chromosome is a single loop of DNA. Bacteria also possess extra bits of DNA on small rounds called *plasmids* that are central to the DNA trade. Like modular storage units, plasmids contain 20 or 30 "auxiliary genes" that encode biological toxins, enzymes enabling the digestion of novel food, or antibiotic resistance, among other things. When bacteria reproduce, just as the single chromosome is copied and passed on to offspring, plasmids can be passed from parent to daughter. But here is where things get weird. While we humans hold tight to our genetic stock, passing it like a carefully tended trust fund vertically from one generation to the next, bacteria pass plasmids from one cell to another like day traders on the stock-exchange floor.[26] In a process of so-called bacterial sex, when bacteria are in close proximity, a hair-like thread of cell membrane extends from one cell to another forming a "conjugal bridge." Plasmid DNA is transferred *horizontally* between bacterial cells. In environments like our guts, crowded with bacteria, plasmids can be passed around like juicy bits of gossip. Even more bizarre, plasmids can pass between different *types* of bacteria. Unlike us, bacteria share genes with siblings, friends, and neighbors. This horizontal gene transfer provides bacteria with an unimaginably deep and interconnected gene pool.[27] And *that* is a concept with which we are just now coming to grips.

If that thought isn't enough to make us sit up straight and vow to take antibiotics only when absolutely necessary, consider the recent findings by Kiran Bhullar, Gerry Wright, and their colleagues. Deep in New Mexico's Lechuguilla Cave is a place that has been isolated from the outside world for over 4 million years, safe from all of our chemical mayhem, including antibiotics. Yet when Bhullar and others collected bacteria from deep within the cave, they found resistance to a shopping list of antibiotics. "Like surface microbes," they write, ". . . some strains were resistant to 14 different commercially available antibiotics."[28] One of those genes confers resistance to daptomycin, an antibiotic of last resort for patients suffering from MRSA. That the gene has existed for millions of years is humbling. It is as if the joke is on us. This collection of genes coding for resistance is now referred to as the *resistome.* Yet studies also show that some pathogenic bacteria isolated before the large-scale use of antibiotics *lack* resistance genes even in their plasmids. The lack of historical resistance in staph, strep, and myriad other pathogens, combined with the provenance of the resistome, raises an obvious question: Why does the resistome exist?

Recall that the majority of antibiotics in circulation today did not originate through human invention but rather through human discovery. Like penicillin, many antibiotics are derived from chemicals that we've co-opted and then spread around the world in unprecedented quantities. Some may well be chemical-warfare agents evolved long ago by biota engaged in perpetual conflict. It's not difficult to imagine fungi in nature like penicillin sparring with bacteria over limited resources.

But other antibiotics, suggests Julian Davies, may simply be part of a chemical messenger system.[29] Life has a long history of signaling from one cell to another, and small antibiotic-like molecules are their language.[30] Once a message has been sent and received, a cell would do well to destroy it rather than allowing such messages to build up like so much chemical noise, or worse, as constant stimulants. Consider the neurotransmitter acetylcholine, which signals many of our muscles to contract. Continuous stimulation can be lethal. Organophosphate pesticides kill by inhibiting acetylcholine breakdown. Our cells have evolved plenty of enzymes devoted to chewing up and destroying chemical messages. It makes sense that bacterial cells, too, carry genes for chewing up and spitting out naturally occurring antibiotics—no matter their role in nature. As Bhullar and colleagues write, there is increasing evidence that these nonpathogenic bacteria provide "a reservoir of resistance genes."

Our dependence on antibiotics, combined with the promiscuity of plasmids, is moving resistance into pathogenic bacterial populations. Even synthetic antibiotics based on novel mechanisms do not guarantee victory. There too, bacteria can make an end run. Consider ciprofloxacin, a powerful second-generation synthetic antibiotic, once the answer to resistance. Pathogenic bacteria are now resisting even this powerful drug; one strategy, explains Julian Davies, is by co-opting an *existing* enzyme that has deactivated antibiotics with *no* structural similarity to cipro.[31] This is like using a wrench to do the job of a screwdriver—a testament to nature's ingenuity.

But wait, there's more: as resistance circulates around hospitals, communities, and local farms, some geneticists suggest that antibiotic overuse may even increase the evolutionary *potential* of bacteria—their "evolvability."[32] Bacteria were already masters of evolution, but we may have made them even better. Our use of antibiotics has given pathogenic bacteria no other option but to evolve into superbugs; and we are inching backward toward the days when disease triumphed and families hid in their homes, away from diseased neighbors, wary of catching their death. How do we avoid this post-antibiotic-age scenario?

The Bigger Picture

Put simply, we must reduce our evolutionary footprint. Of the 37 million pounds of antibiotics consumed in the United States, about 7 million pounds are tossed down our throats or injected directly into our veins. A whopping 30 million pounds are fed to pigs, chickens, and cows. The emergence and spread of antibiotic resistance in a society that didn't grasp the power and the nuances of evolution is, perhaps, forgivable. But that is no longer the case. We are all responsible in one way or another, and we must all contribute to the solution.

"We have been using antibiotics in agriculture since the 1940s—that means enormous amount of use in nonhumans," says Davies, who is on a mission to educate the public. "In India, 10 times as many infections are antibiotic resistant—because antibiotics are available over the counter."This is the case in various other countries as well. National news outlets like Canada's *Globe and Mail*, says Davies, often headline lethal food-borne infections. "What they forget is that hundreds of people die because of *antibiotic resistance*. But it's the sensational items that get the press." Despite two back-to-back editorials on resistance in the *Globe and Mail*, Davies says it will take far more to instill fear of resistance into readers, writing to the paper's editor: "I urge you to publicize the threat of antibiotic resistant infections. What about using space on the sports pages with photographs of amputees as a result of infections with flesh-eating disease? A few tables showing deaths from such infections in Canadian hospitals every week or so would also be helpful."[33]

Davies' frustration is understandable. There was a time when amoxicillin ruled in my own home. While the kids were young, the drug was as common in our refrigerator as milk, juice, and eggs. If it wasn't

Today, there are indications that physicians, at least, are getting the message. When asked how my young son should treat the weekly turf rash burned into his arms and torso—an all-too-common affliction of high school and college athletes in these days of artificial turfs, his pediatrician suggests forgoing the antibiotic cream for soap and water. Now when patients arrive showing symptoms of a possible viral illness, such as a cold or flu, instead of treating immediately, the CDC recommends an "antibiotic timeout" as physicians wait for diagnosis even though culturing for pathogen confirmation can take a day or two.[38] Had I dragged my sniffly kids into the doctor's office today, most likely I'd be leaving the office empty-handed.

But patients are only part of the problem. When 80 percent of antibiotics in the United States are consumed by farm animals primarily for growth promotion, the role of agriculture in pathogenic resistance becomes undeniable. In 2013, the United States took a step forward by urging "judicious use" of pesticides in food animals, encouraging a phase-out of antibiotics for production. So why not simply ban the non-therapeutic use of antibiotics in farm animals? To paraphrase a 2009 report from the American Academy for Microbiology, it's complicated. "Banning the use of antibiotics for growth promotion," they write, "has been shown to increase the use of antibiotics later in the animal's life for treating infection." Information on the impact of agriculture in the medical clinic, they say, "is scant."[39] Meanwhile Srinivasan says that while agricultural use likely impacts human health, "finding direct studies is a challenge."[40] And industry argues that the problem is complex, citing reports like that of the Academy.

But when several dozen consumers are sickened by ground turkey contaminated with salmonella resistant to the antibiotics that might have cured them, as happened in 2008, the link between antibiotic use in agriculture and human health is difficult to ignore. The European Union banned antibiotics in animal feed back in 2006; reports suggest subsequent reductions of some resistant bacteria in livestock.[41] Still, the US Food and Drug Administration remains resistant to aggressive regulations of agricultural antibiotic use. Rather than following the EU's lead, the FDA instead asks that veterinary drug companies voluntarily take charge, re-labeling drugs so that those important for human use are reserved for the prevention, control, and treatment of infection.[42] It is a strategy some within the agency question.[43] So while many physicians around the country are now more judicious when prescribing antibiotics, the prognosis on the feedlot is unclear.[44]

New antibiotics are also essential. At the outset of the twentieth century, Ehrlich's and Domagk's laboratories relied on trial and error as they tested one chemical after another. Today's drug developers can comb through mounds of genetic data, seek out vulnerable targets, identify potentially effective new antibiotics, and rapidly screen tens of thousands of chemical compounds. They know more about how antibiotics work, about how resistance evolves, and about bacterial genetics. Large-scale studies of genes and proteins, combined with analysis of the evolutionary histories of genes and gene networks, can help researchers identify weaknesses in the bacterial armor, ensure the target is unique to bacteria, and find targets that likely require several mutations to evolve.

So where are all the new drugs? Despite all the technological advances, new antibiotics have yet to appear at a steady clip on pharmacy shelves. In contrast to many other drugs, antibiotics tend to be highly complex molecules. Developing synthetic drugs as effective as chemicals tested and selected over the course of millions of years is a little like setting uninitiated soldiers down in the middle of a millennia-old tribal conflict and expecting them to defuse the situation.

When antibiotic discovery peaked in the 1950s, six of the major classes were derived from natural products, while only one originated from a synthetic molecule. Since then only four new classes have emerged, three derived from natural products and one synthetic.[45] And for each new antibiotic, a resistance gene is, even as I write, working its way around the hospital floor, or a crowded apartment building, or a high school locker room. For a pharmaceutical industry bent on developing and selling blockbuster drugs, antibiotics seem to offer little in the way of financial reward. Drugs for chronic ailments of the heart, blood, or brain are profitable. Drugs against rapidly evolving pathogens, though, make poor candidates for cash cows. Fortunately, there are still riches to be found in nature, and our small pool of naturally derived antibiotics, writes the Cold Spring group, are only the "tip of the iceberg." But harvesting nature's magic bullets is no longer the work of a single insightful chemist or microbiologist—drug discovery today requires a spirit of collaboration between academia, government, small biotech, *and* a pharmaceutical industry that is understandably skittish about spending time and money on drugs that could be rendered useless in a few years.[46]

We are circling back toward the pre-antibiotic age. As Maggie and other health-care workers around the globe struggle to hold the line against pathogenic bacteria, we must alter course. In today's highly traveled and far more populous world—when infection can spread around the globe in a day—prevention and cure are more critical than ever. In the fall of 2012, Srinivasan hoped to capture the public's attention when he declared: "The threat of untreatable infections is real. Although previously unthinkable, the day when antibiotics don't work is upon us."[47] In 2014, a World Health Organization report reiterated the sentiment, stating that antibiotic resistance around the world has reached alarming levels.[48] But it is not like we haven't been warned before. First there was Fleming. Decades later another Nobel Prize winner, Joshua Lederberg, similarly warned that "we live in evolutionary competition with microbes. . . . There is no guarantee that we will be the survivors."[49]

We carry around a huge load of bacteria, many of which may shape our lives in ways we have yet to understand. Indiscriminately killing all of them while aiming at only one or two no longer makes sense. The modern age of discovery has laid bare the power of evolution and provided insight into its inner workings. We now know that no matter how many new antibiotics we discover, there will always be resistance. We must rebalance our relationship with the world of microbes—pathogens, essential and nonessential bacterial—perhaps someday even pitting our resident bacteria against disease-causing bacteria.[50] The evolution of resistance is inevitable but its *pace* is not. We have imposed powerful selection pressures. It's time to discover a new way: a way to save the patient without killing the antibiotic.

CHAPTER 2

Prevention: Searching for a Universal Vaccine

"Get the shot," urged Annie. Her good friend and fellow physician K. was considering forgoing the yearly flu shot, even though it could mean losing her job at the hospital. K. was having health problems of her own, and like many chronic sufferers of undiagnosed illnesses, she had begun to wonder if all our Western medication was doing more harm than good.

K. isn't alone in her reluctance. Messages concerning the safety and efficacy of flu vaccines fill our in-boxes, news channels, and magazines. In 2010, roughly 20 percent of physicians and nurses opted out. Many unvaccinated health-care workers worried about the side effects of vaccination or were concerned that it might make them sick.[1] Yet an increasing number of clinics and hospitals are making flu vaccination a requirement of employment. So just how sick was sick? And what about the doctors' pledge to protect others?

Each year in the United States, the flu virus is associated with anywhere from 3,000 to nearly 50,000 deaths, depending on the strain. While the elderly are typically the most vulnerable, the flu also takes too many of our youngest, particularly infants, who are more likely to be hospitalized. Recently, the CDC revised the number of estimated deaths associated with the 2009 H1N1 flu from 18,500 to hundreds of thousands, and possibly half a million, deaths worldwide.[2] In 2013, a lethal flu virus, H7N9, never before identified in humans, emerged in China; within a year it had infected over 200, killing 65. As world health organizations work to develop a vaccine in case the virus begins spreading easily from person to person and causes a pandemic, yet another lethal flu virus new to humans, H10N8, has begun circulating in China.

When flu is on the rise, so too is the quantity of information and misinformation spread about in coffee shops, school cafeterias, or online. Kids tell their friends that the vaccine is lethal, and websites warn of government conspiracies. Aside from the more dire claims, the prevailing assumption is that you can get sick from the vaccine. But flu shots contain only pieces of the virus or inactivated virus and cannot cause flu. And the nasal vaccine contains a weakened form of the virus that may cause some mild and brief discomfort, though nothing like full-blown flu. It is true that vaccinated individuals may end up with the flu; perhaps they were exposed within the two-week window it takes the vaccine to become fully effective, or maybe the vaccine was not 100 percent effective against a particular flu, or they may not have mounted a strong response. The flu vaccine is the least likely cause of illness. *I* know this. Yet, while driving my sixteen-year-old down to the pediatrician's for a set of booster vaccines in addition to the annual flu shot, I couldn't shake the feeling I was subjecting her beautiful healthy body to a medical intervention that may not even be necessary—and may have adverse consequences I cannot begin to imagine. I might read and write articles about influenza vaccines, but when it comes to my own children, I pause. Rationally, I know vaccinations pose little risk and are some of our best bets for disease prevention. But I also know as a scientist that we tend to test for a limited set of outcomes based on whatever knowledge we have at the time.

Flu is a slippery beast. There will always be a new flu, because the viruses have particularly high rates of mutation and because the human immune system is incredibly good at what it does, which puts the pressure on flu viruses to evolve—naturally. This relationship between virus and immune response is ancient. Once flu takes hold it can spread around the world in a blink of the eye. Antivirals may help calm the storm, but they too are often bested by rapidly evolving influenza. For those who are most susceptible, prevention can be a lifesaver. But if immunity drives virus evolution, could vaccines influence evolution as well? And if that is the case, is there a way to use this to our advantage? Could we someday outwit influenza and disrupt this age-old evolutionary game?

Flu in View

"I have the flu" has become a catchall for anything from a stomach bug to a cold to the real deal. It is only when we are *really* struck by the flu that we learn the distinction. Actual influenza is not a little queasiness in the gut, or the low fever that we tolerate as we go about our jobs. Flu can knock us off our feet. The worst flu viruses kill us. The infamous Spanish flu pandemic of 1918 is estimated to have taken 50 million lives worldwide, possibly killing more soldiers during World War I than did the bullets and bombs. Attacking a disproportionate number of healthy adults, it was a medical mystery. The great influenza must have been incredibly humbling for scientists who thought they were on the brink of preventing infectious disease. As researchers discovered one disease-causing bacterium after another, other illnesses, including yellow fever, polio, smallpox, and flu remained frustratingly elusive. Decades after bacterial pathogens had come into view, viruses continued to slip like water through medical science's finest filters.

The discovery that influenza could be transferred *from* the filtered nasal drippings and throat "garglings" of human researchers—who had incidentally caught the flu—*to* laboratory animals in 1933 was a step forward. A more sophisticated understanding came two years later when chemist Wendell Stanley finally captured viruses in crystalline form. The filterable agents that had caused so much death and destruction were, as it turned out, not much more than a bit of genetic material wrapped in protein.[3] They weren't cellular like bacteria, but they did contain genetic material—the essence of life as we know it. It was known by then that viruses could replicate just like any living cell. And they could mutate. But they were astonishingly small. Were they alive or inanimate? Stanley, who won a Nobel Prize for his work, wrote that despite our tendency to "regard them as very small living organisms," viruses curiously "overlapped with the organisms of the biologist at one extreme and with the molecules of the chemist at the other."[4]

Seventy years later, the nature of viruses remains a mystery. We know that some viruses cause us to sniffle and sneeze while others kill within hours. We know there are a billion viruses in a teaspoon of ocean water and trillions living within us.[5] And we now know that throughout our existence viruses have woven in and out of life—leaving their stamp on most if not all living things, perhaps even contributing to evolution's fits and starts. By some accounts, up to 8 percent of our genetic material came to us by way of viruses. Virologist and molecular biologist Louis Villarreal has observed, "The huge population of viruses, combined with their rapid rates of replication and mutation, makes them the world's leading source of genetic innovation. . . ."[6] It is enough to give one pause. We are so intent on stamping out pathogenic viruses and disease-causing microorganisms, but many of these microbes have made us who we are today.

That said, we needn't worry about being "too successful" in our campaign to eradicate viruses; we've managed to wipe only one human viral disease, smallpox, from the planet. Even as our kids receive nearly a dozen recommended vaccines and boosters before the age of two, targeted diseases persist.[7] Viruses continue to break through barriers thrown up by vaccination programs—whether because of the fallout of political struggles, distrust of vaccines, misinformation (like the now-discredited link between the measles-mumps-rubella vaccine and autism),[8] or because of complacency (no one seems to get polio, mumps, measles anymore—so why vaccinate?), or because we are simply beaten by the evolutionary process. And one of the most frustrating viruses for vaccine developers is the flu. Making vaccine is relatively easy. Making a vaccine against the prevailing flu virus is a test of our ability to detect, identify, and predict the unpredictable—evolution.

Evolving Rapidly

In 2009, a flu called H1N1 jumped from pigs to humans. By the time the World Health Organization declared a public-health emergency, flu had sickened more than a thousand in Mexico, killed more than a hundred, and had made its way north with 20 confirmed cases in the United States. Panic set in. Mexican citizens visiting China were treated as if they had the plague and were forced into isolation, Israelis were advised against Torah kissing, and the French were reconsidering "La Bise." The flu had not only evolved the ability to jump from pigs to humans (much to the pork lobby's chagrin as they protested the "swine flu" label), but it had also evolved the ability to transmit from human to human, and it was spreading—rapidly. Initial cases were relatively mild, but there was always the potential for it to quickly evolve into a killer, as others had done before it. Consider the HIV virus that causes AIDs or the inevitable novel influenzas that scare the pants off of scientists working in disease prevention and control. Like bacteria, viruses experience high mutation rates, strong selection pressures, and rapid generation times, creating the perfect evolutionary storm. Mutation is the key to life's diversity, but too much of a good thing can lead to disaster—if not for the virus, then for their hosts.

Just as mutation is essential for evolution, so too is maintaining some level of DNA integrity; otherwise a genome would spiral out of existence. Where there is DNA replication, whether viral or human, there are DNA-repair enzymes. Like my spellchecker flagging typos, a molecular proofreader checks for mistakes as DNA replicates. Before it adds another DNA building block, it looks back to see that the last block added was correct. If incorrect, it deletes that block and adds the correct one. Even so, mistakes happen. To protect against these errors, mismatch-repair enzymes correct the newly synthesized DNA. Bats, bedbugs, and bacteria all evolve through mutation while maintaining species integrity through repair. Even DNA viruses like herpes and smallpox viruses benefit from repair and proofing enzymes. But flu is an RNA virus.

Typically RNA is copied from a DNA template before it is "translated" into protein. In the normal scheme of things, if the DNA is in good shape there is little need to recheck and correct the RNA. When RNA is the template for replication, as it is for flu, mutations can accumulate more easily. RNA viruses are far more mutable than DNA viruses or DNA-based life in general, and the RNA mutation rate can be astoundingly high. Columbia University professor of immunology and microbiology Vincent Racaniello has observed that, given the mutation rate and size of the polio viral genome (an RNA virus), an infected cell could theoretically produce *10,000* new viral mutants.[9] At that rate, it is no wonder that RNA viruses like the flu are so highly evolvable.

Even so, not all flu viruses are equally hazardous to our health. There are three types of influenza virus with varying capacity to make us sick: influenza A, B, or C. The flu that causes us to line up for vaccines, compulsively wash our hands, or take to our beds is influenza A or B. On the other hand, we are likely to mistake influenza C for a cold. Both A and B are relatively simple, each comprised of eight RNA segments carrying the necessary information for building new viruses. Encoded in these segments are instructions for two influenza proteins that are the bane of our existence: hemagglutinin (HA) and neuraminidase (NA). Because they don't have all the bits necessary to reproduce on their own, viruses *need* growing cells—or more specifically, a healthy cell's machinery. Once flu virus finds its way to its target cell, HA acts like a socialite's calling card allowing the virus to slip its RNA into the cell without causing undo harm. Should I become infected, my respiratory cells will quickly become subservient to the virus. Like a Xerox copier gone berserk, my own cells will churn out thousands to millions of copies of viral RNA. New viruses will be assembled and packaged, ready to be sent out into the world. This is where NA comes in. Just as HA enabled the entry, NA allows the next-generation viruses to be released from the cell's membrane. As new viruses burst forth, my respiratory cells will be left in tatters.

Many of these new viruses will be replicas of their parent. But others will be mutants: perhaps their HA calling cards are no longer recognizable, or their NA proteins are defective, or they are so lethal that they kill before a body has the chance to sneeze out millions of viral progeny. Some mutations, however, enable the virus to carry on by evading the immune response. If I become infected, my immune system will fight back, eventually producing antibodies that, like predator drones, seek out and attack viral proteins. Most likely, they will attack HA and NA. However, it is also likely that by the time my immune system becomes fully armed, some of the invading viral army will have already mutated. A slight change in the HA protein will be enough to make it unrecognizable, giving the virus a chance to thrive again. If I pass this virus on to my kids, husband, or neighbor, I will inadvertently contribute to this particular virus's evolutionary history. As the infection gains traction in my husband or neighbor, the cycle repeats. This is called *antigenic drift* (*antigen* referring to the bits of virus protein that our immune system recognizes). Endless antigenic drift is the main reason why those of us who choose to do so must line up for a new flu vaccine each year.

Over the course of the flu season, public-health scientists track infections, noting the minute changes in HA and NA proteins. Tracing flu evolution in this way is essential for vaccine reformulation. It's sort of like predicting fashion trends for the upcoming season, except that in this case being caught out of style costs lives. Usually the annual vaccine targets two prevailing strains of influenza A and an influenza B. Antigenic drift keeps public-health officials and vaccine makers on their toes, but drift alone does not produce the kind of flu that keeps them awake at night. Every once in a while flu makes an evolutionary leap, and when it does the outcome can be devastating.

Influenza is native to birds. But over the past century, three subtype combinations have flourished as "human influenza": H1N1, H3N2, H2N2 (numbers denote different serotypes of HA or NA). These are the viruses that evolved the ability to not only flourish in our throats and lungs but also transmit from me to you. These human viruses tend to prefer humans, just as bird viruses specialize in birds and pig viruses prefer pigs and so on. Jumping from one species to another is like throwing a saltwater fish into a freshwater pond; a few might have the means to survive, but most won't. So the cat that cuddles with its sneezing, sniffling human stays healthy (and vice versa). But every once in a while, influenza A jumps from one species to another. I may catch flu from my dog; a teenager may become infected by her prized pig; or, in one of the more bizarre scenarios, a whale may catch bird-flu by way of its blowhole. These are the sporadic cases we hear about in the news. When vendors or shoppers at an Asian poultry market come down with bird flu, or H5N1, we hear about it because flu jumping from bird or pig to human can be catastrophic. Of the 622 cases of H5N1 reported over the past decade, 371 were fatal.[10] Should H5N1 evolve the ability to pass through a sneeze or a cough, we may find ourselves bracing for pandemic flu. This is the kind of virus evolution that sparks the imagination of moviemakers and which keeps health organizations on edge. Many fear that another 1918-like flu is only a matter of time and that, with our penchant for global travel, it will be far more deadly.

While a transmissible version of H5N1 has yet to evolve "in the wild," the virus recently shot to infamy (as did several researchers) when it evolved the ability to transmit from one ferret, a model for human influenza, to another. Researchers in two different laboratories had driven the virus to evolve into the kind of flu that causes pandemics. Their goal was to identify mutations and biological characteristics that would make H5N1 transmissible, should it evolve "naturally." The existence of a transmissible subtype of a potentially dangerous virus—even if in ferrets (where it caused mild flu)—was enough to instigate a frenzy of doomsday articles, concerns about biosecurity, talk of censorship, emergency meetings, and research moratoriums. For a while, the research was too hot to publish.[11] At the heart of this evolutionary brouhaha in one case was a series of four mutations in HA, possibly making way for a very dangerous liaison. Despite the controversy, those mutations will provide virus trackers with a potentially life-saving heads-up. But these kinds of major shifts in a virus's character don't always happen one step at a time.

Sometimes influenza A takes a giant evolutionary step, known as an *antigenic shift.* Picture the eight RNA segments as a deck of colored cards. A green deck represents the genome of a strain that infects birds, and a blue deck represents the genome of a strain that infects people. Should they both infect a common host, the decks get shuffled together when the viruses reproduce: the green deck can pick up some blue cards and vice versa. Should the bird virus acquire a human influenza HA, it will carry the blue calling card typical of human flu, enabling it to infect people. This is how flu variants make large evolutionary leaps. "All the human flu we know," says Cornell virologist Colin Parrish, "are at least partially derived from . . . one strain," referring to an H1N1strain that has evolved through a century of antigenic drifting and shifting.[12]

lected and purified; attenuated virus is harvested to use in nasal sprays.

Over the years, I have watched as my children's immune systems were poked and prodded with a multitude of vaccines: inactivated tetanus and diphtheria toxin; bits of pertussis bacteria; and various attenuated viruses including those causing polio, chicken pox, measles, and influenza. By the age of six they had been vaccinated against 10 diseases and received at least twice that many vaccinations, including booster shots. This may be "normal"—but it certainly is not "natural." Could this antigenic barrage on the young immune system push the system beyond its capacity or alter its natural course? Even as I wondered about the potential consequences, I offered up the arms of my little ones and watched as they were swabbed, poked, and bandaged. The concern about vaccination overload is common enough that the Centers for Disease Control has responded by posting a note on their website. There is no evidence, they say, that "recommended childhood vaccines can 'overload' the immune system." The CDC notes that from the time babies enter this world they are essentially swimming in a sea of antigens, from the bacteria living on their bodies and found in their food to the antigens clinging to their hands and other objects that find their way into a baby's mouth "hundreds of times every hour." And of course babies are bound to be exposed to pathogens. A cold virus might present just a handful of antigens, while strep exposes one to dozens.[17] An article published over a decade ago in the journal *Pediatrics* reports that children are exposed to far fewer antigens per vaccine than their parents. When I was injected with the smallpox vaccine decades ago, some 200 different proteins dispersed into my bloodstream; today the whole complement of 11 recommended vaccines contains fewer than 130 proteins.[18] Our capacity to respond is indeed extraordinary.

So it seems we can handle the invasion—but what of the invaders? How might they respond to this great wall of immunity? Viruses, says infectious-disease expert Andrew Read, may be no different than any other organisms engaged in an ongoing predator–prey relationship, demonstrating a wide variety of responses to selection pressures. One of Read's many interests is the evolution of virulence in *response* to vaccination. "Obviously," says Read, "vaccines are strong selection pressures."[19] A virus might become invisible to the immune system, or replicate faster, or shift from one preferred target tissue to another. For over a decade, Read has focused on Marek's disease in poultry—a highly contagious tumor-forming virus. Chicken farmers have traditionally protected their flocks through large-scale vaccination. But over the years, as viruses acquired immunity to vaccination, the disease seems to have gained virulence, killing chickens more surely and swiftly. Could vaccination have caused Marek's virus to evolve into a more effective killer? "What stops super nasty bugs from circulating in the world?" asks Read. "The usual answer is that if they kill the host, they kill themselves. So then, what happens when we keep the host alive with a vaccine?" Read's research suggests that keeping a host alive may allow more-virulent viruses to evolve and survive, like Marek's. It is an idea, Read says, that when first published in 2001 was extraordinarily controversial, and it still is.

But there is a precedent for increasing virulence in response to a host's resistance, and the story of the *Myxoma virus* in rabbits is a textbook example. In the mid-1800s, desiring a "spot of hunting," an English settler named Thomas Austin imported and released two dozen rabbits on the grounds of his Australian home.[20] He was not the first, nor the last, to release the nonnative rabbits, but Austin's stock in particular is credited with the ensuing rabbit infestation that has plagued the country ever since. Lacking natural predators, the vigorous breeders become an unnatural disaster within a decade—their incessant grazing blamed for reductions in both plant and animal biodiversity. No amount of mass slaughter could put an end to the infestation. Australia's famous "rabbit-proof fence" was another famous failure. By 1907, after six years of construction, over 2,000 miles of wire fencing stretched across the evolution seems confined to HA and NA. With so few proteins (recall the deck of just eight cards), each with an essential role, flu may have less wiggle room for the emergence of dramatically new characteristics. Changes to HA or NA could possibly be associated with enhanced virulence, enabling quicker entry or exit from a cell, but further research is needed to evaluate these kinds of changes. Even so, this relatively simple little virus has us scrambling: testing, predicting, and creating new vaccine, every year, year after year. We now know a good deal about how influenza evolves, what drives it to evolve, and which antigens on which proteins evolve, all necessary ingredients to undermining influenza's evolutionary escape route.

One Vaccine to Rule Them All

"We tend to make vaccines that mimic the natural viruses that are circulating, so that they induce the same immune responses as those viruses," says Colin Parrish. "But we are doing it to a virus that has evolved for hundreds of millions of years. Trying to recreate the natural response is not likely to be the most effective"—particularly if we want a broadly acting vaccine that, like many childhood vaccines, retains efficacy well beyond a year or two. Doing so, as Parrish suggests, means that vaccine developers must consider throwing the virus a curve ball by identifying and aiming for antigens that are unlikely to undergo evolution or that differ in subtle ways from the natural viruses—so that the virus has less experience at escaping the response.

If there is a Holy Grail for influenza control and vaccine research, it is a universal vaccine. This catchall vaccine would confer immunity regardless of the virus strain. Rather than provoking an immune response toward the usual suspects like HA, a vaccine might instead be based on less variable proteins—proteins that change little over time and that evolve far more slowly than HA or NA. There's promise on this front. According to immunologist Suzanne Epstein, some candidate vaccines against these highly conserved antigens "are effective in animals against all influenza A strains that have been studied."[23] Another strategy is to direct the immune response to a region of the HA protein that has changed little over time. Like a mushroom, HA protein has a globular cap and a stem. The cap is a common target for our immune response and evolves rapidly, but the stem is somewhat "hidden" from attack. As such, it is one of those regions that changes little. Recent studies suggest that if our immune system can be prompted to make antibodies against the stem, we may be able to fend off different strains of influenza. In other words, rather than producing antibodies effective against a single strain of influenza, we could produce antibodies that are "broadly neutralizing," although generally only for one of the two groups of influenza A.[24] Epstein says there are unanswered questions as to the efficacy of universal vaccines in comparison to existing vaccines, and as to how influenza virus might evolve in the presence of these different kinds of immunity. Some effects could be a concern, while others might actually slow viral evolution.[25]

It is difficult to predict the impact of a universal vaccine against a notoriously evolvable virus. And a universal vaccine may not free us just yet from regular vaccines—but it may at least extend the time between vaccines. It might also protect us from unknown and newly emergent influenza A viruses, reducing the likelihood of a 1918-like pandemic. Longer-lasting vaccines, combined with greater trust in vaccines, may throw a wrench into evolution's engine by making the transmission and therefore the distribution and spread of evolved virus more difficult. If enough individuals are vaccinated, breaking the chain of infection, then this sort of "herd immunity" may also protect those who are not vaccinated or who fail to mount a sufficient immune response. In essence, says Katia Koelle, "the idea would be for the vaccine to lower incidence and to simultaneously slow antigenic evolution (which would further lower incidence)."

How and *if* universal vaccines will influence the evolution of targeted proteins is unpredictable. But without a universal vaccine, influenza virus will undoubtedly continue its rapid evolution under pressure from our immune response. There are always trade-offs, and we do the best we can with the information we have. In this case, we have the potential to protect vulnerable populations—our elderly, our youngest, and those who are immunocompromised—against a virus that is notoriously unpredictable and highly evolvable. No one knows how influenza will respond to universal or broadly neutralizing vaccines. Nor are vaccine developers promising an evolution-proof vaccine but simply one that would slow the process. Given our current position in the race against influenza, we could use the head start.

If there is one thing that I hope to make clear with this book, it is that evolution is inevitable. It is a process in which we are intimately involved—naturally, usually inadvertently, sometimes intentionally. We have engaged with influenza virus for millions of years; our immune system is one outcome. Universal vaccines may provide the opportunity for us to be a little lighter on our toes, perhaps reducing our evolutionary footprint on influenza viruses. At the very least, a universal vaccine may buy us valuable time, allowing an escape from the vaccine treadmill.

As I write in the waning days of summer, flu season is on its way. Whatever influenza viruses are circulating, they will have evolved over the past year. Maybe the change will be small. Or maybe it will be large. Flu will infect young and old, spreading through day-care centers, retirement homes, airports, and grocery stores. The dispute over vaccines will once again crest and spill over to local papers and other news media, blogs, coffee shops, schools, and the workplace. Last year, Annie's hospital granted K. a waiver. This year the hospital has instituted a firmer policy: waivers will be less likely. Flu is predictably unpredictable. So too is our immune system's ability to triumph on its own. Relying on natural immunity might work for the human species as a whole, as it has for millennia—there will be survivors. But it may not be enough for vulnerable individual members of our communities: the infant down the street, the asthmatic next door, or the elderly grandmother across town. New vaccines may allow us to slow the evolutionary process, giving us all a fighting chance. This may well be the ounce of protection that is worth far more than the pound of cure.

CHAPTER 3

Treatment: Beyond Chemotherapy

"When I was diagnosed with chronic myeloid leukemia cancer, my wife and I were both runners, training pretty high miles," recalls Matt W., a 37-year-old physicist and father of three. "At the time I hadn't noticed anything in particular was wrong. I had a few episodes of night sweats, but I didn't put it all together." But when Matt could barely finish a 5K race, it was clearly time to see a doctor. As a graduate student, he didn't have a regular physician, so he went to a walk-in clinic. "The poor doctor, who had never seen me before, basically said 'you've got white cells off the charts and your spleen is the size of a grapefruit.'" The diagnosis was leukemia. That was Friday. Matt was scheduled for the oncologist the following Monday, which "gave me and Elizabeth [Matt's wife] all weekend to freak out about it."

It was 2003, and a new drug, Gleevec (or imatinib), had just received FDA approval as a first-line treatment for the disease. Studies at the time showed it worked—at least for a while. Gleevec controlled the advancement of chronic myeloid leukemia (CML) from the less aggressive chronic phase to the more dangerous accelerated phase or final blast phase, until for some patients, it didn't. But it was far better than a bone marrow transplant. For a healthy candidate with a good match, transplant survival rates hovered around 70–80 percent. Matt went on Gleevec and would consider a transplant if resistance arose. Matt and Elizabeth had been married for under a year. Kids were in the more distant plan. "When I was diagnosed, there were maybe seven known healthy babies and no sick ones born to men on Gleevec. But no one had any idea what would happen three or five years out if Gleevec didn't work. I'd need a transplant and after that you can't have kids, so we decided to just start a family." Jack was born about a year after Matt's initial diagnosis.

Just a few years earlier, before the discovery of Gleevec, Matt might have been told he had five years at best without a stem cell transplant. Gleevec dramatically changed the odds of survival for *most* CML patients. Chronic myeloid leukemia is a cancer of the white blood cells. Normally, the concentration of circulating white cell counts hover around 3,000–10,000 cells per microliter. When he became sick, Matt's white cells were in the hundreds of thousands. At some point, they broke free from the checks that keep one cell type or another from growing out of control. Gleevec (and other similar targeted agents) kill the aberrant cells by inhibiting the overactive enzyme at the root of the disease, returning counts to the normal range. Its success was unprecedented. Unfortunately, Gleevac doesn't work for all patients; their disease is either unresponsive from the beginning or eventually becomes so. And, though Matt had a good initial response, would his CML—whether in a year, or five, or ten—evolve resistance?

For much of the twentieth century, cancer treatment tended to be more palliative than curative, particularly chemical therapy or chemotherapy ("chemo"). But cancer is a disease governed by basic evolutionary principles—including the remarkable and heritable variation of cancerous tumors and the natural response to intense selection pressure imposed by chemotherapy. A group of scientists who recently convened to explore chemically induced resistance, "regardless of the biological system," refers to cancer as an "evolutionary disease."[1] To break the cycle of treatment failure, we must place both cancer and cancer treatment in an evolutionary context.

The Seeds of Cancer

Eventually, one way or another, we are all touched by cancer. A minority of us are predisposed, whether to breast, colon, or skin cancer, our cells encoded with DNA that make them more likely to break the rules and grow wildly out of control. Some cancers, for one reason or another, may be promoted by diet, or perhaps by environmental toxicants or infections, or some inscrutable combination of all of the above. For the majority, an underlying cause has not been found. Whatever the primary etiology, chances are good that cancer will someday take root in me or in someone I love. Perhaps it already has. Each year, more than 1,600,000 Americans will be diagnosed with some form of cancer; and each year, cancer will be listed as the underlying cause of death on hundreds of thousands of death certificates. Many more will remain ignorant of the cancerous cells within their bodies; or they will live a perfectly normal life with low-grade cancers.

Cancer mirrors the drawbacks and the benefits of our technological age. The disease has plagued humans since the dawn of our time. Yet by most accounts, it is more prominent today than at any other point in our history. Perhaps this increased incidence is a trade-off for lives lived longer, combined with our dramatically altered lifestyles and industrial-age environmental changes.[2] Exposures to natural and synthetic chemicals through air, water, fruits, and vegetables, the paint on our walls, the beds we sleep in, and even the clothes we wear are unprecedented. Too many of us still smoke cigarettes—one of the more significant chemical exposures. Our current chemical environment, diet, lifestyle, and perhaps most importantly, prolonged life span, all combine to provide ample opportunity for the enterprising cell to break away. At the same time, recent technological advances in detection and diagnosis add to our awareness of cancer's pervasiveness and improve the odds of survival.

Cancer, of course, is not unique to humans but is a disease that afflicts all manner of multicellular life. Clams, fish, and frogs can get cancer, and even dinosaurs were susceptible.[3] A body—whether it belongs to a mosquito or a mouse—is like a society of cooperative and interdependent cells that have evolved a sort of order. There are collaborators, independents, watchdogs, and renegades. Cells are in constant communication; immune cells roam throughout; chemical messages flow from one cell to another; limbs and organs emerge in a fetus almost perfectly every time. These things happen through highly conserved processes—the outcome of more than a billion years of evolution. And all parts of these complex systems retain the capacity to evolve. Otherwise, we wouldn't be here. But when key genes encoding controlling proteins and enzymes are hobbled or altered, a cell may buck the body's *social organization*: perhaps there are 11 toes rather than 10; or an enzyme fails to metabolize an essential nutrient; or a cell transforms from collaborator to renegade—enabling a breast cell to break free and invade the liver, the bones, or the brain.[4] Like yin and yang, cancer is a natural outcome of a tightly controlled but essentially evolvable system subject to some degree of disorder.

As with any living things (as well as viruses), when cells divide and DNA replicates, mistakes happen. The overwhelming majority of these mistakes will be corrected. Those that persist remain in our extensive genome as mutations. Most are of little consequence. A few may be lethal for the cell. Some confer a survival or growth advantage. This kind of variation is essential for evolution. When we talk about evolution of humans and other animals, we are concerned with the changes that occur in our *germ* cells—those residing in eggs and sperm—which can be passed on to our offspring. These changes are potentially heritable from one generation to the next. When we talk about the evolution of cancer, most often we're referring to mutation and evolution that occurs in our *somatic* cells—breast cells, liver cells, or brain cells—the cells that make up our bodies. In the latter case, the impact of mutation is on the individual affected cell, its progeny, and eventually the entire organism. At least one scientist has suggested that a cancerous cell is like an "animal within"—a parasite that has evolved from our otherwise healthy cells.[5] When cancer arises, a subset of renegade cells prospers at the expense of the greater society of cells that comprise our bodies.

In most cases, a single mutation does not cause cancer, nor do all genes have the potential to contribute equally, or at all, to the evolution of cancer. When altered, certain genes—usually those essential for cell growth and survival—will send a cell down a potentially destructive path. Some of these "cancer genes," or oncogenes, are familiar, like the BRCA1 and BRCA2 genes associated with breast cancers; others less so, like the KIT gene that can lead to cancers along the digestive tract. Cancer genes are estimated to comprise 1.6 percent of our genome.[6] As I ponder the tens of trillions of cells in my body and the roughly 21,000 genes comprising our human genome, the opportunity for cells to take flight is ever present. But, just as our immune system holds the line against pathogenic bacteria, viruses, and even cancers, our genome has its share of control freaks—genes and proteins that constantly check, fix, and even terminate those processes deemed abnormal. The aptly named "caretaker genes," for example, encode proteins and enzymes that repair and maintain genetic integrity, while "gatekeeper" genes keep in check the cells that have managed to slip by. The products of these genes monitor cell communication, growth regulation, and death.

When our surveillance system fails, or itself succumbs to pathogenic alteration, conditions become ripe for cancer.[7] But even then, cancer is not a simple process. Several years ago, molecular oncologists Douglas Hanahan and Robert Weinberg proposed a set of six steps that cells undergo as they are reprogrammed from "normal" cells into cancerous cells: they must acquire self-sufficiency in growth signals, become insensitive to signals that would normally inhibit growth, evade programmed cell death, acquire limitless potential for replication, sustain growth of blood vessels, and be able to invade other tissues and spread throughout a body. All together, they wrote, these alterations "represent the successful breaching of an anticancer mechanism hardwired into cells and tissues."[8] While the inactivation of a single caretaker gene may be insufficient to cause cancer, the genetic instability that follows can open the door for other mutations, freeing cancerous cells from the "social" constraints of normal cells.[9] When cancerous cells do arise, the local environment—tissue density, blood flow, nutritional status—may also influence its capacity to flourish.[10] Yet despite all the checks and balances and environmental caveats, cancer arises often enough that it kills one out of every four Americans.

Almost as common as cancer are the books and articles pointing to lifestyle and dietary changes that can reduce cancer incidence; we have made great strides reining in the release of carcinogenic chemicals, thus reducing (though certainly not eliminating) exposures from air, water, food, and consumer goods. Still, cancer remains with us. Depending on the cancer, treatment options, including surgery, radiation, immunotherapy and chemotherapy, are continually improving. When the tumor is "fluid" like leukemia, a bone marrow transplant is also an aggressive form of anticancer therapy. But in many cases, surgery isn't an option and transplant can only go so far, and moreover both carry risks of their own. So the question arises: How can we selectively eradicate disease cells without killing the patient? And, as with antibiotics, can this be done through targeted therapy?

Mortal Combat

A cancer patient diagnosed today stands a better chance of surviving the disease, if not living essentially "cancer-free," than at any other time in history. The industrial age has been certainly associated with an increased cancer incidence, but it has also ushered in a golden age of cancer treatment. Ancient healers and physicians, when confronted with tumorous growths including malignancies, turned to herbal potions, fermented fruits, and poisonous metals. Boils were lanced, blood was let, tumors excised and cauterized. If a patient didn't die from such treatments, success was fleeting at best. Cancer, unlike many infectious diseases, is a progressive disease (and some cancers are *caused by* pathogens, like herpes virus).[11] There are plenty of individuals who, without treatment, survive infectious diseases, but a cancer patient will seldom beat the disease without treatment. When antibiotics and vaccines first radically altered our relationship with infectious diseases, cancer remained essentially untreatable. Antibiotics were effective and relatively nontoxic because they interfered with proteins and processes unique to bacteria. Cancer presented a greater challenge: differentiating an abnormal human cell from a normal human cell. Cancers that did form discreet tumors might be surgically removed, but those that coursed through veins and arteries, like Matt's CML, were initially even more difficult to control.

One important difference between a cancer cell and the normal precursor cell from which it is derived is its life history: birth, death, reproduction, and all the activities in between. Cells in different tissues have different life histories according to their respective roles in the body. Skin cells protect us from the outside environment, whether temperature, sunlight, toxic chemicals, or physical damage. We lose skin cells all the time, and so they are constantly dividing—turning over in a matter of weeks. The cells lining our guts—one of our body's harshest environments—might slough off in a matter of days, and so, like skin cells, they are constantly replenished. Other cells, like muscle, bones, and nerves, may live for months or years.[12] White blood cells are more like skin and gut cells. They are constantly produced and may live for days rather than weeks or months, which means that the bone marrow—where blood cells are produced—contains plenty of actively dividing cells. Cancers, having lost some degree of growth control, may also have large populations of cells that are in the process of cell division. Whereas our muscles and brains may be like an old-age home, our blood and guts are more like multigenerational housing—teeming with youngsters. And this provides oncologists with a target. A chemical that kills dividing cells will kill cancer cells preferentially over most other cells. But it may also kill off other actively dividing cells, including gut cells, the skin cells in our hair follicles, and white blood cells. Chemotherapy agents, particularly the earlier agents, are simply cytotoxic (cell-killing) chemicals.

One of the first effective chemotherapy agents, not surprisingly, was valued not for its *curative* properties but for its efficacy as a killer chemical. We know this chemical today as a notorious agent of war—mustard gas. Deployed by the German Empire during the First World War on the battlefields of Europe, most infamously in Ypres, Belgium, mustard gas—a relatively simple combination of sulfur, carbon, and chlorine—killed hundreds of thousands of French and colonial troops. Over a million others were sickened and maimed for life. Once it made its way into the body, the chemical also affected tissues with larger proportions of dividing cells. Wartime autopsies found the lymph nodes, spleens, and bone marrow of victims depleted of white cells. It was a curious finding, but, as nations tried to put the horrors of warfare behind them—banning the use of chemical weapons in 1925—any follow-up studies with mustard gas were kept under wraps.[13]

Less than two decades later, with the rise of Nazi Germany and the onset of the Second World War, the threat of gas warfare lingered despite the ban. Mustard gas may have been "gone" from the battlefield, but it was by no means forgotten—which ostensibly explains why, in 1943, the American Liberty ship *John Harvey* was carrying a load of mustard gas bombs. The bombs were intended for retaliation, just in case the Germans reneged on the treaty. Docked in the old port city of Bari, Italy, the cargo likely would have slipped through the war and evaded the history books had the Germans not raided the port. On December 2, as German planes bombarded Bari, sinking 28 cargo ships including the *John Harvey*, nearly 100,000 pounds of mustard gas spilled across the harbor and rose into the night sky. Thousands of soldiers and citizens were exposed. Hundreds were hospitalized with chemical burns and blindness. At least 83 died. The cause was a mystery to all but a few "in the know." Upon autopsy, it was found that the victims' white-blood-cell counts were oddly depleted.

By the time of the Bari incident, leukemia was fairly well characterized as a cancer of the white blood cells. And secretive studies into the effects of mustard-gas-derived chemicals on white blood cells were beginning to bear fruit. Experiments by pioneering pharmacologists Alfred Gilman and Louis Goodman revealed astonishing efficacy of one mustard-like chemical that targeted white blood cells in laboratory mice afflicted with lymphoma. Typically, laboratory mice with lymphoma lived about 21 days. The first mouse treated with the mustard agent lived a remarkable 84 days.[14] After two doses its tumor regressed. The chemical seemed to target cancerous white blood cells. What Goodman and Gilman couldn't have known then was how the mustard derivative worked—why it seemed to target white cells and not most others. Years later, studies revealed that the chemical slips into the DNA molecule, rendering it incapable of normal replication. Ultimately, the hobbled cells die. Since it targets cells in the process of replicating—those that reproduce most often, including cancerous white blood cells, are preferentially killed. Unfortunately, the chemical's efficacy was fleeting. Cancer cells, observed Gilman, were remarkably resilient. When dosing stopped, the cancer bounced back. Worse, it became increasingly tolerant to drug exposure. Yet, even though cancer control was short-lived, the ability to melt away a tumor through *chemical* treatment was unprecedented. In 1942, the first human subject suffering from an advanced leukemia was injected with nitrogen mustard. The response, writes Gilman, "was as dramatic as that of the first mouse."[15] Exposure to the mustard-gas derivative had chased the cancer into remission within days. However, as with the mice, disease respite was temporary.

The significance of such a breakthrough was not unlike the discovery that infectious disease could be cured by antibiotics. But despite the promising results, advancing chemotherapy in the 1940s was a battle. Gilman and Goodman's early research remained secret until after the war. In addition, wrote Gilman, "in the minds of most physicians the administration of drugs other than an analgesic, in the treatment of malignant disease, was the act of a charlatan."[16] One can only imagine how a pharmacologist who injected leukemia patients with a chemical-warfare agent would have fared. After the war, though, prejudices began to drop by the wayside. Although further studies confirmed mustard's efficacy, it was not a miracle drug. Not all patients responded in the same way; the drug was so toxic that treatment had to be carefully balanced with morbidity and lethality; and inevitably, treatment failed due to therapeutic resistance.[17] Still, chemotherapy derived from mustard gas and other chemicals granted cancer patients a reprieve from death: a few weeks, months, or years—sometimes long enough for the next new drug.

A new age of cancer treatment had begun. Chemicals from mustard-gas derivatives to analogs of folic acid, a B vitamin, were employed in the war against cancer.[18] Still, drug discovery and development remained an expensive and tedious process, with more error than trial. Writes hematologist and author Guy Faguet in his bracingly candid book *The War on Cancer: An Anatomy of Failure and a Blueprint for the Future*, "The majority of cancer drugs today exist because drug developers either got lucky or were incredibly persistent—testing one chemical after another."[19] For nearly 60 years, chemotherapy has essentially targeted rapidly dividing cells without distinguishing between cancerous and normal. Unlike antibiotics, which exploit the differences between bacteria and our own cells, chemotherapy drugs kill our own cells, often to devastating effect. Treatment regimens depend, in part, on how much collateral damage a patient can bear. How many healthy cells can be poisoned without killing the patient? Even when a suitable balance is obtained, too often it doesn't last; resistant cancer cells proliferate and doses escalate beyond what is tolerable.[20] When placed in an evolutionary context, the failure of chemotherapy becomes clear. Just as antibiotics impose strong selective pressure on pathogens, chemotherapy imposes incredibly strong selection pressures on cancer cells. Cancer cells are already on the evolutionary fast track, and so chemotherapy selects for resistant cancer cells time and again. It is no wonder that even with highly effective treatments like Gleevec, it is difficult to shake the fear of resistance.

Resistance Rising

The emergence of chemotherapeutic resistance is a phenomenon with which oncologists are all too familiar. Life evolved in a toxic world—so it is no surprise that many strategies for surviving toxic environments are highly conserved, passed from one generation to the next, retained across species. A cell may protect itself by metabolizing and transforming toxic chemicals into relatively harmless by-products; or by preventing them from entering in the first place; or by sequestering or excreting noxious chemicals. These time-tested strategies have enabled life to flourish in ever-changing environments fraught with toxic chemical threats. And sometimes a cell carries a slightly altered version of a gene that might incidentally provide protection—a variation on a theme that might just happen to come in handy. A population of cancer cells will likely contain some cells better equipped for survival than others, which may explain one of the more vexing strategies for resistance in cancer cells, baseline genetic variation.

It is not necessary that genes conferring resistance to a treatment always arise anew, as we might tend to think. Standing genetic variation offers fast-track evolutionary change in all manner of living things from bacteria to birds, bees, and—as more recently discovered—cancer cells. Perhaps a gene that happens to confer resistance is already present in a population of cancer cells. Like a sweater stowed away in the corner of the wardrobe that suddenly comes into style, a gene might be tucked away in a cancer cell's genome and come into play when toxic chemicals flood into its environs. Sometimes resistance genes exist in a handful of cancer cells well before exposure to chemical treatment, making a cancer resistant from the get-go; this is primary resistance. Dr. Amir Fathi, an academic oncologist at Massachusetts General Hospital and Harvard Medical School, explains that primary resistance during therapy is seen when patients fail to meet established response benchmarks: "They just fail to get to the point where you'd like them to get, so there is a suggestion of resistance from the beginning. Then there is secondary resistance—patients may respond initially, they may have a very robust response," but, says Fathi, over time the response is lost.[21]

In Matt's case (Fathi was not his oncologist), Gleevec, combined with another drug, hydroxyurea, helped return his blood count to normal in a little over a month. Matt began running again. But the search for a bone marrow match continued, just in case, as oncologists monitored Matt's blood each week for signs of resistance. Despite the positive signs, back in 2003 CML was still known as a disease with a propensity for resistance. Whether Gleevac would continue working was not yet clear.

Bad Blood

Chronic myeloid leukemia is just one of several blood cancers. According to American Cancer Society estimates for 2013, about 6,000 individuals in the United States will be diagnosed with CML; about 600 men and women will die from it (though this number is dropping as therapies improve).[22] Blood is our body's information highway. Coursing through our veins are red blood cells, white blood cells, clot-forming platelets, and myriad proteins and chemicals—nutrients, waste products, and messages that are passed among cells and organs—carried along in blood plasma. When white blood cells turn cancerous, the balance between white and red blood cells tips in favor of white cells. As concentrations of red cells and platelets are reduced, skin bruises more easily, less energy becomes available. A 5K race turns from a sprint to a slog. In the early stages of CML, the white blood cells are still functional. They may crowd out other blood components, but they retain the ability to function as white blood cells. But they are not "normal."

The vast majority of CML cells carry an oddity known as the "Philadelphia chromosome," in which a stretch of DNA is swapped between two different chromosomes—reminiscent of the genes that are swapped like cards between viruses, but in this case very specific pieces of chromosome are exchanged. When the exchanged bits recombine, a new functional gene is created. Located on the Philadelphia chromosome, the new gene, called BCR-ABL, codes for an aberrant form of an enzyme called tyrosine kinase. Tyrosine kinases are critical for normal cell function, regulating the activity of proteins by essentially signaling them either to become active or to cease activity. A tyrosine kinase that is out of control is like a disruptive kid at a birthday party who's had too much cake and ice cream. As she spins out of control, she sets off every other kid—and chaos ensues. The BCR-ABL kinase is out of control, stuck in the "on" position, and so cells with the Philadelphia chromosome grow out of control. Within the context of the hallmarks of cancer, CML is a one-hit kind of model. The altered chromosome seems to provide CML with all it needs to grow; eventually a young woman notices she bruises more easily or that she's experiencing night sweats, or a runner like Matt finds himself uncharacteristically exhausted at the end of a short race.

Early on, the goal of chemotherapy was to slow CML's inexorable progression toward a "blast phase." Unlike the early stage of CML, when white cells maintain their function, cells produced in blast phase are often much more dysfunctional and can accrue additional mutations. Many fail to mature and fail to function as normal white blood cells. Extreme fatigue, increased susceptibility to infection, and uncontrollable bleeding may ensue. It is the least treatable phase of the disease. Prior to 2001, if a newly diagnosed patient was lucky, they might have been eligible for bone marrow transplantation—a procedure fraught with its own complications, even in the best candidates. If not, their lives may have been prolonged with cytotoxic chemotherapies such as busulfan. Like nitrogen mustards, busulfan throws a wrench into DNA replication. But busulfan doesn't distinguish between cancer cells and normal cells. And, like other chemicals that interfere with DNA replication, including nitrogen mustards, busulfan can cause cancer in its own right (a side effect of chemicals that interfere with DNA replication is that, in the process, they may cause mutation).[23] Hydroxyurea is also used to treat CML, as it was for Matt, specifically to reduce the cell count during the initial phases of treatment. The chemical slows DNA replication by interfering with the production of key molecules required for DNA synthesis. Prior to the emergence of Gleevec and similar targeted agents, it could push average survival of CML patients out to five years.[24] Ten years before Matt's diagnosis, in the early 1990s, the authors of one study bluntly observed that "survival of chronic CML is not much better today than 70 years ago. The principal mode of treatment has been palliative chemotherapy. . . ."[25] Chemotherapy didn't cure—it just helped treat the symptoms. The imposition of a powerful selection pressure enabled the emergence of resistance; in response, chemotherapy doses increased. Eventually, the treatment itself became intolerable.

Interferon-alpha—a chemical normally produced by our own immune system—offered some respite and extended survival rates. Depending on the stage of the disease, the drug could push average survival times beyond 10 years in a subset of patients.[26] But interferon came with its own slate of side effects, from flu-like symptoms to depression and thyroid dysfunction. And cancerous cells eventually evolved resistance. "The growing dogma," wrote Siddhartha Mukherjee of that time, "was that CML was perhaps intrinsically a chemotherapy-resistant disease. . . . By the time the disease was identified in full bloom in real patients, it had accumulated a host of additional mutations, creating a genetic tornado. . . ."[27]

Like Matt, Doug Jenson was an athlete—a runner and cyclist. But, when Jenson was diagnosed in 1997, treatment options were limited to chemotherapy and transplant; he was given three to five years to live.[28] Just as his body began to reject the interferon treatments, Jenson enrolled in a clinical trial for a very different kind of cancer drug called imatinib, developed by Oregon Health and Science University researcher Brian Druker. Unlike chemotherapy, imatinib didn't kill indiscriminately or interfere with DNA but targeted CML's out-of-control tyrosine kinase. As one of the first effective targeted therapies, imatinib, marketed as Gleevec, was revolutionary. Like a heat-seeking missile, it zeroed in on and inhibited the altered BCR-ABL tyrosine kinase. If the kinase flipped a cell's "on" switch, Gleevec gummed up the enzyme, incapacitating it. Healthy cells don't have BCR-ABL, so collateral damage was minimized. The few side effects tended to be far more tolerable. Gleevec was like chemotherapy's penicillin moment.

Nearly 10 years after switching to Gleevec, Doug Jenson—describing himself as "healthy, happy, and active"—celebrated his 50th wedding anniversary.[29] Mukherjee writes of his own Gleevac moment that, when examining the blood of a patient on the drug in 2002, "not a single leukemic blast was to be seen. If this man had CML, he was in a remission so deep that the disease had virtually vanished from site."[30] When Matt's oncologist confirmed CML, he said, "This is great, this is fantastic! We've got this new drug and we're going to just fix you up." Matt thought the guy was crazy. The pattern of resistance to Gleevec had yet to show up in clinical studies that were available to Matt as he combed the online literature, awaiting his first meeting with the oncologist. Based on the available studies, said Matt, "You thought you'd get on Gleevec and it would last two or five or ten years and then you'd develop a mutation and Gleevec wouldn't be very effective."

Matt is one of the more than 80 percent of patients whose CML remains in check with little evidence of resistance. For a small proportion of CML-positive patients, it is but another stop-gap measure.[31] Within five years of treatment, resistance will arise in about 14 percent of CML patients. Of the resistance strategies "available" to cells, CML cells seem to make use of them all: a mutation that changes the BCR-ABL protein, prohibiting Gleevec from combining properly with the protein; multiple genes producing multiple copies of the tyrosine kinase, swamping out the drug's beneficial effects; cells equipped with pumps better able to block Gleevec's entry. So far, more than 100 different mutations conferring resistance with varying degree of influence have been identified.[32] Says Marco Gerlinger, an oncologist and researcher keenly interested in cancer evolution, several studies on the origins of imatinib resistance in CML suggest that many patients carry mutations conferring resistance from the get-go, before treatment even begins. In those cases, Gleevac simply provided the selection pressure kicking the evolutionary process into gear.[33] "This is quite worrying, as it shows that treatment-resistant cancer cells are already present in many patients at the time we make the diagnosis," adds Gerlinger.

Fortunately, thanks to a better understanding of the common paths of resistance evolution, there are now emerging generations of alternatives to Gleevec that may manage resistance effectively by addressing them from the beginning. There are fewer side effects as well. Even for CML with a mutation called T315I, which has conferred resistance to almost all previous targeted agents, a new drug, ponatinib, holds promise. "It's been a desperate search for something that would address these patients," says Amir Fathi. "Because they don't respond to imatinib, dasatinib, nilotinib—name your 'ib.'" Patients with highly resistant mutations have needed stem cell transplants, since their disease is a challenge to get under control, explains Fathi. Just approved by the FDA, ponatinib may hold promise for highly resistant forms of CML too, providing long-term remissions in previously challenging clinical scenarios.

CML treatment is a success story. But for the majority of cancers, a highly effective targeted treatment has yet to be discovered. "The majority of tumors," says Gerlinger, whose practice currently focuses on advanced kidney and prostate cancers, "develop drug resistance after several months on therapy. . . . The reasons why the majority of cancers become resistant is one of the most important questions in oncology." Understanding why Gleevac and other targeted CML therapies are comparatively resistance-resistant will likely open the door to all manner of targeted treatments—perhaps even, beyond cancer. Is it the nature of the drug? Or the cancerous target? "Simplistically," says Gerlinger, "we think Gleevec works better than many other therapies, because CML is essentially driven by a single driver which can be effectively inhibited. Other cancers have multiple drivers and we can only inactivate very few of them, so some of the hallmarks of cancer remain active." Given the challenges and successes, then, how might oncologists extend the efficacy of treatment? How can they help their patients step off the resistance treadmill?

Taking the Evolutionary Road

Carlo Maley directs the Center for Evolution and Cancer at the Helen Diller Family Comprehensive Cancer Center in San Francisco. In a review titled *Overlooking Evolution*, Maley and colleagues, including lead author Athena Aktipis, suggest that evolutionary approaches provide "unrealized opportunit[ies]" for advances in cancer therapy research. As with highly resistant CML, one of the greatest stumbling blocks to cancer treatment is standing genetic variation. "Every known cancer drug," writes Aktipis and colleagues, "suffers from this problem and it is the primary reason we have not been able to cure cancer."[34] Adds Aktipis, "Virtually all cancer deaths in the developing world are due to therapeutically resistant disease."[35]

Decades ago, cancer was treated as if it progressed from one mutation to another, morphing from normal to cancerous like a phrase in a game of "Telephone." This idea of clonal growth dominated cancer research and treatment strategies for years. If cancer emerged from a single clone, then a single biopsy would yield sufficient information about the whole tumor or the cancer's progression. But, says Gerlinger, "the 'monoclonal origin of cancer' misled many oncologists and cancer researchers," who believed that cancer-cell populations remained monoclonal as they expanded.[36] In reality tumors are more likely made up of subclones, each following their own evolutionary trajectory, like branches on the tree of life. An evolving population of cells will eventually bud out with new mutations, and some branches will end up in a dead end. Others branches will flourish. And as with any evolutionary scenario, diversity within a population depends, in part, upon mutation rate and population size.

Cancer-cell lineages expand under pressures exerted by their local environments, which may explain why some cancers have a higher mutation load than others. Melanomas and lung tumors in smokers, for example, are influenced by sunlight and cigarette smoke, respectively. Both may have hundreds of *unique* mutations and there is evidence that lung cancer in smokers may have more mutations than lung cancers in nonsmokers. A cell is particularly prone to spawning mutants if DNA-repair genes mutate. And some cancers may become more heterogeneous as we age. Certain types of cancerous cells that arise in children may have less opportunity to acquire mutations than, say, the cells in my middle-aged body. All told, mutations in a detectable tumor can number in the billions (although only a subset of those may contribute to cancer growth and survival).[37] All of this mutation and adaptation has prompted cancer researcher Robert Gillies and colleagues to liken tumors to "continents" that are "populated by multiple cellular species that adapt to regional variations in environmental selection forces."[38] If resistant subclones constitute a minority of cancer cells prior to treatment, a biopsy may easily miss them, making it difficult to predict how a patient might respond to treatment. But, says Gerlinger, things are changing. More oncologists have started to figure out how to detect and monitor the many different subclones. "My impression," he says, "is that a paradigm shift is on the way."[39]

So where does cancer treatment go from here? How does all this evolutionary know-how help the patient? When the best route is either chemotherapy or targeted therapy, how can oncologists delay or even prevent resistance? One strategy is to *use* evolution. A treatment might keep cancer in check by allowing for "deselection." When the selection pressure is reduced, *drug-sensitive* cells might once again dominate the population.[40] The cancer may not be gone—but it may be manageable. Cancer cells might also be led into an "evolutionary trap": if resistance to one chemical therapy leaves the cancer cell more vulnerable to another, then one counterintuitive strategy would be to actually encourage the evolution of resistance.[41]

Knowledge of evolutionary processes can give oncologists an edge against cancer. "Cancer cells can only adapt to immediate selection forces," write researchers Robert Gillies, Robert Gatenby, and Daniel Verduzco. "They cannot anticipate future environmental conditions or evolutionary dynamics."[42] But we can anticipate, and *that* plays to our advantage. Our strategy toward cancer, they write, ought to be more "chess" than "whack-a-mole."[43] A few years ago, Gatenby proposed a provocative and novel approach: treatment, but with restraint. "Instead of focusing exclusively on a glorious victory," writes Gatenby, practitioners might instead consider ". . . the possible benefits of an uneasy stalemate."[44] Rather than aiming to eradicate cancer, treatment would strive to maintain a tolerable but susceptible population of cells. Sort of like suggesting that rather than strive toward excellence, a population instead maintain some mediocrity—because in this case, the path to excellence might just lead to resistance. Writes Gatenby, "In battles against cancer, magic bullets may not exist, and evolution dictates the rules of engagement."[45] Gerlinger agrees:

> We need to understand much better how cancers evolve and what fosters the evolution of aggressive and resistant clones. Approaches that aim to prevent this by maintaining a population of slow-growing cancer cells in order to prevent the outgrowth of more-aggressive cancer clones are extremely important. . . . Finding the Achilles heel of drug-resistant cancer cells may allow us to develop combination therapies that prevent the outgrowth of resistant cancer-cell clones from the start.[46]

These strategies, no doubt, will require a dramatic shift in how we think about treatment. Is it possible to make a sort of peace with such an aggressive and lethal opponent?[47]

Fortunately for Matt, these aren't yet questions that must be answered. His CML is now effectively undetectable, even at the molecular level, and it has been for years. Even so, treatment strategies may change as the science progresses. Recently, a European group published the results from a study in which patients with undetectable disease had gone drug-free for at least two years—albeit with mixed results. Some stayed disease-free. In others, when signs of CML reemerged, cells were still responsive to BCR-BL inhibitors such as Gleevec.[48] When Matt's oncologist mentioned that some patients had gone off Gleevec, Matt posed the obvious hypothetical question: should he go off it, too? The response, says Matt, was a resounding "No, no, no, no, no, no!" Too many unknowns remain. Why mess with a good thing? What if the disease came back in one of the more advanced stages? Or, what if cells emerged that had evolved resistance? "Seems like it's not worth it at all, right?" says Matt. "As far as I am concerned, I could take Gleevec indefinitely." Or at least until CML is better understood, and the elusive resistance-proof drug is discovered. Then, perhaps, treatment will prevail over one of the most powerful forces of nature—evolution.

CHAPTER 4

Defiance: Rounding Up Resistance

Some call them "superweeds." Weed scientist and anti-resistance zealot Mike Owen calls them "driver weeds." No matter what you call them, agricultural weeds around the world are evolving herbicide resistance, a problem that some claim is one of today's greatest threats to agriculture. "If you don't control them," insists Owen, "you will have a serious economic problem."[1] Herbicide-resistant weeds are gaining ground, and conventional farmers across the country are scrambling for solutions. The weeds that get Owen traveling across the country and around the world, appealing to farmers, academics, industry scientists, and ag management alike, aren't just any resistant weeds. These are the horseweed, pigweed, waterhemp, and others that have evolved a stubborn resistance to Roundup, the "once in a century" herbicide. Resistant weeds, hundreds of different species, are spreading across the country at an increasingly rapid pace, infesting more than half of our nation's crops. By the time you read this, their numbers will have increased.

As much as I would prefer my food and clothing to be produced herbicide-free, I am sometimes lured by the lower cost and availability of early-season, conventionally grown strawberries, brilliant red peppers, and ripe avocados. The reality is that our family eats plenty of products grown with herbicides, as do the great majority of consumers in the United States, and Roundup has surely left its mark on our diet. If I had to pick a poison, I'd want it to be the least-harmful for my family and the environment. Roundup's active ingredient, glyphosate, is touted as the less-toxic alternative to herbicides like 2,4-D and atrazine (although recent studies are calling its safety into question).[2] Since the 1970s, Roundup has been the go-to herbicide for relatively low toxicity, but its popularity spiked when Monsanto developed the first-ever herbicide-resistant crop—Roundup Ready soy. The combination created a cropping system. You buy the Roundup Ready seeds and you use the Roundup herbicide. The technology made it easy for growers to use Roundup anytime, anywhere. The crops, says Owen, changed the entire face of agriculture. As farmers across the country adopted resistant crops, Roundup became the number-one herbicide in the country. But, like a rising Hollywood starlet, the herbicide was loved to death. Says Owen, "When Roundup Ready technology became available, growers basically jumped on it like a duck on a June bug."

Roundup Ready crops dominate farms in the United States, which means that most of us are unwittingly engaged in a marriage of convenience with engineered crops and with Roundup. Two of the top six food crops in the country are soy and corn, and the majority are ready for Roundup. So too are the weeds. Since Roundup's introduction, one weed species after another has evolved resistance. As a result, the products I buy may carry higher residues of Roundup, because growers must use more to beat back weeds, or perhaps the products contain a hint of 2,4-D, because the growers have resorted to other herbicides. Or they may cost a bit more. We may be far from the farm, but we are all affected by resistance in subtle ways, whether it's a different cocktail of herbicide residues in our food and water, or the price of a cotton T-shirt. Resistance looms so large that in the summer of 2012, the

National Academy of Sciences—the premier scientific organization the United States—gathered together the top agricultural scientists for a summit on herbicide resistance, with an emphasis on preventing the "once in a century" herbicide from going the way of penicillin. In today's high-tech world, the notion that weedy plants pose a serious threat to humanity seems inconceivable, yet as with viruses, bacteria, and insects, it's not just about the individual species. Evolution of resistance and its impact on our lives is a manifestation of our modern chemical addiction.

This chapter isn't about whether genetic engineering is humanity's savior or downfall; there are plenty of books, blogs, and articles on the topic. Nor is this about *how* weeds have evolved resistance mechanisms, although some background on weeds and resistance is in order. Instead, this is about the consequences of denial. Monsanto built an agricultural house of cards; human nature and rapid evolution are bringing it down. Is it too late to shore it up?

A Weed by Any Other Name

My father was a meticulous gardener. Anything that didn't belong in his garden was extricated with the twist of a knife. Those were the weeds. Whatever remained weren't. He reminded my sisters and me that weeds, by definition, where just that—weeds by definition. A plant with the audacity to grow out of place, whether in our flower gardens, on the organic farm down the road, or at the conventional dairy farm the next town over, is a weed. As we grab and twist, we are more likely to curse rather than admire the traits that enable weeds to flourish in our gardens, lawns, roadsides, and sidewalk cracks. "Weeds are the scourge of the organic farms," says my neighbor, the organic farmer. In 1965, evolutionary ecologist H. G. Baker listed a set of weed characteristics that remain in play today: no special requirements for germination, rapid seedling growth, rapid time to flowering, continuous seed production, the ability to self-pollinate, high seed output, adaptability.[3] In other words, weeds breed fast and often and are not picky when it comes to putting down roots. They are plants that are rapid-evolution ready.

Domesticated crops, on the other hand, are nearly the opposite. Selected by humans over the course of centuries to produce what we want when we want it, crops, like pets, are ill-suited for life in the wild. "Artificial selection," notes biologist E. O. Wilson, "has always been a tradeoff between the genetic creation of traits desired by human beings and an unintended but inevitable genetic weakness in the face of natural enemies."[4] Contrary to the natural process of evolution, we select plants that best serve *our* needs and strive to ensure that crops remain the same year after year; we don't want them busting out and evolving on us. The first farmers selected plants for desired traits such as seeds that remain on the stalk long enough for harvest, or fleshy fruits, fat taproots, and high sugar content. We undermine natural selection with our *un*natural selection of crop plants each time we choose one fruit or seed over another. Meanwhile, weeds are free to undergo unbridled evolution, stumbling across the most innovative ways to survive the pressures of agriculture. Some have even managed to co-opt what evolutionary ecologist Fred Gould describes as "the life of luxury" enjoyed by crop seeds as they evolve into crop-seed look-alikes. Disguised as desirable grains and inadvertently mixed in with crop seed at harvest, these plant equivalents of "snowbirds" enjoy a winter protected from the elements.[5] If it weren't simply an outcome of evolution it would be considered ingenious. "Of all the crop pests," writes Gould, "weeds boast the longest recorded history of adapting to agricultural practices." Coddling our agricultural crops has created a radically uneven playing field, and the weeds are making a run for the end zone. Their biggest triumph may be Roundup—the only herbicide once touted by industry as virtually "evolution-proof."

The Herbicide Roundup

The ultimate goal of an herbicide is to target proteins and enzymes unique to plants, leaving all other living things, from bacteria to beneficial bugs and birds, unscathed. It is a tall order, requiring identification and exploitation of key differences between species. Evolutionarily, plants and animals parted ways over a billion years ago. As plants evolved and diverged into ferns, grasses, dandelions, pigweeds, and palm trees, many ancient genes remained intact; these include conserved genes that are not unique to plants but are common to animals too. But plants also developed plenty of plant-specific genes for proteins, enzymes, and pathways. These are the genes targeted by herbicide chemists focused on picking off pigweed while leaving the birds, bees, and soil bacteria unharmed. In theory. But there are always unintended consequences. Just as good bacteria or even our own cells suffer collateral damage following a course of antibiotics or chemotherapy, herbicides, too, inadvertently harm us or our pets or pollinators. Toxicology is rife with examples of "non-target toxicity"; it is what keeps many toxicologists in business. Even so, in comparison to insecticides or rodenticides and other intentional poisons, most herbicides are not acutely toxic to animals. Of all the bottles lining the garage and garden shelves, herbicides in general are not a good choice for those with malicious intentions.

The ideal herbicide would kill a weed, from shoots to roots, persist just long enough to complete the job and then disappear, breaking down into harmless chemicals. Glyphosate isn't exactly that. It persists, but is relatively sequestered, bound up with the soil. Yet comparisons with penicillin are not just hyperbole; glyphosate is as close to the ideal as the chemical industry has come. Penicillin targets enzymes specific to cell-wall construction in bacteria. Our cells, lacking the cell-wall structure, aren't much bothered by penicillin. Glyphosate targets an enzyme known as EPSPS that is unique to plants, bacteria, and fungi, and is required for the production of essential amino acids.[6] By interfering with EPSPS, glyphosate causes plants to starve to death even in times of plenty. Because we acquire these nutrients in our diets, as do other animals, we are immune to glyphosate's EPSPS activity, but that doesn't necessarily mean that the chemical or its formulation (the combination of "active" ingredients like glyphosate and other supposedly "inert" ingredients) is nontoxic. Few chemicals are so specific that they interact with one system only. There is always the potential for non-target toxicity, particularly as use patterns change. One consequence of Roundup's popularity is heavier use, higher residues, and deservedly increased attention; as a result, a few highly controversial studies recently challenged the claim that Roundup is "non-toxic" in humans.[7]

Despite the current focus on the herbicide's toxicity, glyphosate's discovery in 1970 was an undeniable breakthrough. Coming in the wake of Rachel Carson's *Silent Spring*, Roundup—though surely a pesticide—came with the promise of leaving fields devoid of weeds yet ringing with birdsong. Glyphosate was so revolutionary that it landed Monsanto chemist John Franz in the National Inventors Hall of Fame, alongside the innovators responsible for Post-it notes, rockets, plastics, copiers, and birth-control pills. Glyphosate soon became a top-selling herbicide, eventually playing an integral role in the rise of no-till farming, which meant less time on the tractor and better conservation of soil, water, and fuel.[8] By the 1990s, tens of millions of pounds of glyphosate were being applied to roadsides, pastures, and farm fields, and Roundup had become one of the world's best-selling herbicides.

But Monsanto had even bigger plans. The advent of genetic engineering—the art and science of manipulating an organism's genome—enabled plant and animal breeders to do in months or years what had, until then, taken decades or more to accomplish, if accomplished at all. If the natural process of evolution or even evolution under human selection pressure didn't provide, nature could be engineered. In the early 1980s, Monsanto set their sights on creating crops resistant to

Roundup.[9] Because *crops* would be immune to the poison, farmers could use the herbicide virtually anytime and anywhere. If Roundup had been Monsanto's king, resistant crops would be their queen.

Evolution Not

To allay concerns, proponents of genetic engineering point to humanity's long history of forcing evolution in plants and animals. After all, without human selection corn wouldn't be so sweet, lettuce would be bitter, and carrots would be more like Queen Anne's lace, their wild relative. Our agricultural ancestors knew nothing about DNA and genes. They simply selected for emergent favorable traits: less bitter, sweeter, bigger. Still, there is a hint of nature in this sort of nurture. The desired characteristics arose naturally and randomly, an odd mutation here and there ripe for the picking. Genetic engineers do not wait for favorable traits to arise but instead seek out genes for desirable characteristics that may exist in wildly different species and insert them into their target species. In 1972, Nobel Prize–winning biochemist Paul Berg paved the way for this revolutionary and controversial technology when he recombined DNA from two different viruses.[10] Like words cut from one document and pasted into another, genes from one organism could be isolated and inserted into another. Some two decades later, one of the first fruits of biotechnology moved from the greenhouse to the dinner plate when Calgene, an upstart company located in Davis, California, marketed Flavr Savr tomatoes. For those who don't recall, the tomatoes contained a gene extending the shelf-life of naturally ripened tomatoes—an answer to artificially ripened, bullet-like tomatoes that looked nice but were tasteless. Initially a success, engineered tomatoes quickly fell from grace, the victims of poor planning by the producers as well as consumers' concerns over bioengineered foods.[11] Since then, engineering has given us a freakish menagerie: a luminous rabbit engineered to express jellyfish genes; "environmentally friendly" pigs; goats that produce spider's web proteins in their milk; yeast that produce a malaria drug; and a growing number of herbicide-resistant crops.

Journalist Daniel Charles has chronicled the early days of Monsanto's genetic engineering feats in his book *Lords of the Harvest.* In the beginning, writes Charles, bioengineers saw themselves as green "revolutionaries." These were scientists who had come of age during the 1970s and viewed biotechnology, such as it was, as the answer to agriculture's chemical addiction. Fears of genetics gone awry aside, biotech could help rein in the large-scale use of toxic pesticides. A case in point were plants engineered to express Bt toxin, a pesticide swiped from *Bacillus thuringiensis.* Strains of the soil-dwelling bacteria (a relative of anthrax bacteria) secrete toxin lethal to certain species of caterpillar, moth, and nematode pests of crops like cotton, corn, and potato. Eliminating the need for industrial chemicals, the bacteria have been the organic farmer's friend since the 1920s. For bioengineers, identification of the Bt toxin gene was an opportunity to improve upon nature. Isolated in the early 1980s, its discovery, writes Charles, was "too perfect a target for the fledgling biotechnology industry to ignore."[12]

By the end of the decade, the first generation of gene jockeys from Monsanto and elsewhere had figured out how to cut and paste the Bt toxin gene into laboratory plants. Today the crystalline toxin is expressed in corn, cotton, and potato crops grown in the United States and around the globe. Now there is controversy over the potential for toxicity to non-target species, and resistance is evolving in target species. But if pesticide reduction was the goal of Monsanto's early bioengineers, Bt was a shining example. Engineering Roundup resistance was something different, though. The only reason to instill herbicide resistance in plants would be to *encourage* the use of Roundup. It was already a top seller, but creating a single "system" binding seed and herbicide together would mean billions of dollars and a more solid footing for the herbicide in the agricultural economy. The development of Roundup

Ready crops was like a shotgun wedding for the farmers buying into the cropping system. Some of Monsanto's bioengineers, suggests Charles, regarded the project with scorn, seeing it as a potential stain on the environmentally "clean" nature of biotechnology. But as the funding rolled in, they acquiesced.

Like Bt, herbicide-resistant plants would be produced by transferring a resistant bacterial gene into key crops, beginning with soy. But before the engineering could begin, the bioengineers needed a gene that would confer resistance, which meant figuring out *how* glyphosate starved plants to death in the first place. Just as the Bt toxin gene came from bacteria, the source of a resistance gene would also come from bacteria. But in this case with no known resistance genes, engineers had to first encourage evolution of a resistant form of the gene before they could isolate and patch it into plant DNA. As it turned out, the gene, called EPSPS, was critical for survival and highly conserved. It worked well just the way it was, with little room for modification. Any mutation it acquired would be like altering an old family recipe, difficult to do without changing the taste. Nearly a decade of forced mutation and selection was met with marginal success. And then, writes Charles, "nature trumped all of their efforts."[13] Bacteria inhabiting glyphosate contaminated factory waste had evidently evolved resistance, and they had done so naturally. Once the gene was isolated and transferred into plant cells it worked with little ill effect. The world's first herbicide-resistant crop was poised to put down roots.

With the engineering challenge solved, there were still a few more hurdles before the crops could hit the market. In addition to ensuring that the new product was neither pathogenic nor toxic, the company had to demonstrate that Roundup Ready soy would not itself become a pest species by swapping genes with weeds or other crop species, or that it would not grow wildly out of control.[14] Neither USDA nor EPA had the authority or the means to regulate on *how* new glyphosate use patterns might influence the evolution of resistant weeds. Yet in their petition for *non*regulation, Monsanto's regulatory affairs manager Diane Re sought to assuage concern, submitting letters from a number of weed specialists who claimed that Roundup Ready soy would be "unlikely to increase the weediness potential for any other cultivated plant or native wild species."[15]

"Agriculture," writes plant scientist Jonathan Gressel, "is the largest evolution laboratory presently on earth today, with herbicides as the most ubiquitous man-made artificial selector for evolution on the planet."[16] By the time Monsanto petitioned for nonregulated status, chemical resistance, whether in insects, bacteria, or weeds, was commonplace. In sugarcane fields and along roadsides, resistant weeds had taken hold just a little over 10 years after the herbicide 2,4-D went to market. Today, 30 different species are resistant to the herbicide. There are over 100 different herbicides and over 200 resistant weed species. Water hemp, a native broad-leaved plant resistant to multiple herbicides, has become nearly intractable in the Midwest. An increasing number of weeds now resist more than one herbicide. If we think only about a particular herbicide's target, of the 25 currently targeted sites, weeds have evolved resistance to 21.[17] Plant populations are responding to agriculture's grand experiment in both predictable and bizarre ways: increasing metabolism and excretion of pesticides, ramping up gene copies of the targeted gene (a solution by dilution), and deletion of a single amino acid that changes the resulting protein just enough to confer resistance. Of the latter strategy, writes Gressel, "Never before had nature used that deletion trick . . . not even with evolution of antibiotic resistance in bacteria, not with anticancer or other drugs, and not with the evolution of pesticides."[18]

Despite all the evidence, industry believed that Roundup was different, perhaps because after nearly 20 years there was little evidence of naturally evolved resistance in plants. Several scientists were willing to back up Monsanto's claim. One weed scientist, who had tried and failed to induce evolution and select for Roundup resistance in alfalfa, concluded that "it is unlikely that weeds will develop resistance to Roundup."[19] "The complex manipulations that were required for the development of glyphosate-resistant crops," wrote a team from Monsanto, "are unlikely to be duplicated in nature to evolve glyphosate-resistant weeds."[20] Because geneticists had a difficult time driving rapid evolution of resistance in the laboratory, so the thinking went, it was unlikely to happen in the field. But some scientists who supported non-regulation of the new herbicide suggested that farmers protect themselves from the remote possibility of resistance by using herbicide and crop rotation just as they had in the days prior to Roundup Ready crops. "[It] is my opinion," wrote Mike Owen, "that suggested glyphosate use patterns that would develop as the result of glyphosate-resistant soybeans would not result in the development of a resistant weed population."[21] Another called for the US EPA to "revoke approval of HR crops when and where credible evidence of resistance emerges,"[22] yet there was no federal mandate to monitor or report resistance. And little did anyone know that by1996 Roundup resistance in ryegrass had emerged in Victoria, Australia.

In 1996, the USDA granted Roundup Ready soy "nonregulated" plant status. The herbicide rose from *one* of the top sellers to *the* top seller. Over 180 million pounds were sold in 2007. Roundup had become the most commonly used herbicide in the country. Some industry analysts project that by 2017, worldwide use of glyphosate will rise to over a million metric tons.[23] According to Monsanto's website, the crops "Maximize profit opportunity with no-till" while the providing the farmer with "unsurpassed weed control," taking the worry out of crop production.[24] No matter when the herbicide was applied, resistant crops would stand tall. This meant that farmers could apply Roundup at any time in the growing cycle. But as usage increased, so too did the selection pressure imposed by the herbicide on target weeds. In retrospect, there were bound to be consequences, given Roundup's rising star and the tendency of weeds to grow like weeds.

Monsanto's claim that weed resistance was highly unlikely, combined with the new crop system's ease of use, led farmers to believe they could use the herbicide with impunity. Roundup-resistant soy was just the beginning. Other major crops—corn and cotton in particular but also sugar beets and alfalfa—followed. It was a disaster waiting to happen. One strategy to stave off resistance is to rotate crops and herbicides (with conventional agriculture, each crop comes with its own cast of associated herbicides). But with an increasing variety of Roundup Ready crops, even if crops were rotated, whether corn, cotton, or soy, farmers began applying "Roundup after Roundup after Roundup." "If I could have asked my dad what he thought about the development of glyphosate-resistant crops," said plant molecular biologist Charles Arntzen at a 2012 conference on resistant weeds, "he very likely would have said it was a no-brainer. Glyphosate is environmentally friendly, it gives the farmer greater flexibility . . . and farmers do not have to mix different chemicals. My dad would have adopted glyphosate-resistant crops. Tens of thousands of farmers made that decision."[25]

With the rapid adoption of Roundup Ready crops, Mike Owen's "proper use patterns" never developed. Less than 10 years after their debut, in response to the enormous selection pressure imposed by increased herbicide use, 24 different weed species *independently* evolved Roundup resistance. Over and over again, in different states and different countries around the world, nature outmaneuvered Monsanto's genetic engineers. Weeds survived by way of massive gene amplification (some plants make over 100 copies of the EPSPS gene), by limiting the herbicide's access to plant shoots and roots, and by altering EPSPS enzymes.[26]

Less than 50 years after the discovery of glyphosate, evolution is on the verge of making the "once in a century" herbicide obsolete. With the majority of corn, soy, and cotton grown in the United States, Roundup Ready agriculture will lose its "penicillin" and tens of thousands of farmers will struggle to control resistant weeds. Herbicide resistance, writes the Council for Agricultural Science and Technology, poses "one of the most significant threats to soil conservation since the inception of the USDA Natural Resources Conservation Service."[27] So what is a farmer to do?

There Must Be Some Way Out of Here . . .

"There was a belief that glyphosate would be so difficult for weeds to develop resistance to because of its mechanism," says Mike Owen. "I and others said it would be a problem. For about three years I was *a persona non grata* in St. Louis [home of Monsanto]. . . ." Less than a decade after the introduction of resistant crops, Owen and colleagues organized a National Glyphosate Stewardship Forum aimed at stemming the growth of resistance. The forum included academics, growers, and industry representatives. There was little agreement at that time on the scope of the problem or on possible solutions. The creation of resistant corn, in addition to soy, had reduced any resistance-busting benefits of crop rotation. Instead it was like no rotation at all. "Monsanto strongly promoted the use of glyphosate only; if there had been a modicum of management diversity included with glyphosate and Roundup Ready technology," says Owen, "we wouldn't be in the dire straits we now find ourselves."[28] As with antibiotics, herbicide and crop management is the key to slowing resistance. But Owen, like many other weed scientists, agronomists, and extension specialists, along with Monsanto, Dow, and other major corporations, doesn't promote regulation of engineered crops as a means of reining in the evolution of weed resistance. Resistance says Owen, isn't caused by genetic engineering. To a weed, Owen says, glyphosate is "just another herbicide." It doesn't matter whether it's applied to an engineered or a conventional crop. Growers,

CHAPTER 5

Resurgence: Bedbugs Bite Back

"I was so sleep deprived from worrying and from the itching, I was literally going crazy," recalls Abby of her bout with bedbugs. "The bites were in rows: breakfast, lunch, and dinner. They were terrible bites that itched like nothing before." Desperate for relief, Abby took to sleeping on an air mattress in the middle of the kitchen floor. She'd hoped the problem was fleas, possibly brought in by the cats—at least fleas are easy to kill. But that wasn't the problem. It was bedbugs, which, says Abby, are endemic in town. As a physician and director of a local community health center, she ought to know. She's since seen plenty of bedbug bites and plenty of denial. Abby suspects that, for years, patients with oozy welts caused by bedbugs were misdiagnosed by physicians and nurse practitioners as scabies or flea bites—both easier problems to remedy. The real cause was overlooked for good reason. Like once-common diseases, bedbugs had become part of our history, banished from our homes and apartments decades ago. But now they are back with a vengeance. Pest-management professionals like Orkin and Terminix publish an annual list of top bedbug cities. In 2012, Philadelphia topped the Terminix list, while Chicago topped Orkin's. Reported infestations have risen dramatically over the last two decades. In their 2013 survey *Bugs without Borders*, the National Pest Management Association reports that nearly 100 percent of the pest-management companies surveyed had been called upon to deal with the bugs, up a few percentage points from previous years and far more than a decade ago.[1] And there is no end in sight.

The bugs and their bites are not life-threatening, but an infestation can be crazy-making. And their life history, combined with evolved pesticide resistance, makes extermination notoriously difficult. No exterminator can guarantee permanent eradication, nor can everyone afford the cost of eliminating bedbugs, which can run as high as several thousand dollars. Abby tells of one couple who, upon finding bedbugs in their apartment, tossed their mattress from the second-floor window in disgust. But that only adds to the problem by spreading bugs around the neighborhood. And should we be tempted to pin infestations on socioeconomic class, nationality, or any other "otherness," bedbugs, like fleas and lice, are not afraid to jump cultural barriers. While poorer communities are often hit hardest, largely because they may not be able to afford pricey exterminations, bedbugs have been known to climb to the highest rungs of the socioeconomic ladder. If you sit in a movie theater, go to the doctor's office, buy clothes, attend school, work in an office, stay in a hotel—even New York's upscale Ritz-Carlton[2]—or take public transportation, you just might carry one home clinging to the seat of your pants or the sole of your shoe. "I knew I slept with bedbugs when I woke up and saw an engorged female running away once I turned on the light," says urban entomologist Alvaro Romero of his first encounter with *Cimex lectularius*, in a Kansas City hotel.[3] "Fortunately," says Romero, one of the few next-generation scientists now focused on bedbugs, "I didn't bring them home, but that experience told me that *everybody* is susceptible to bedbugs." The scientist had seen thousands of bedbugs in natural infestations and in the lab, but seeing bedbugs in his *own* bed was another thing.

According to the online bedbug registry, a disturbing number of rentals around my son's university apartment have been infested at one time or other.[4] No surprise there. Look up any college town or city and the registry map will surely light up as students traveling from near and far bring the "gift" that keeps on giving. Some colleges combat infestation by prohibiting students from furnishing their own rooms. Abby suspects that bedbugs entered her home by way of a guest's futon. Just one gravid female, tracked into the home on a shoe, piece of clothing, or suitcase, is enough to set off an infestation. Laying a few eggs a day and as many as five hundred throughout her lifetime, the female produces offspring that breed with one another, all the while feeding upon their slumbering human hosts. Of the rising tide of infestation, one scientist now writes that "until recently most householders and a whole generation of entomologists and pest-control professionals have never seen a bedbug. . . ."[5] The bugs had become so rare in the United States that few scientists bothered to study them. How the bugs slipped back into our high-tech, pest- and pathogen-averse, low-tolerance-for-anything-creepy-crawly society is no real mystery. It is the expected outcome of ever-increasing world travel, demographics, and evolution.

Sleep Tight, Don't Let the Bedbugs Bite

The relationship between bedbugs and humans is as old, if not older, than our relationship with beds. The bugs are thought to have jumped to humans from bats by way of shared cave dwellings.[6] As humans built homes and cities grew large, swaths of the population became infested as a matter of course. The poor crowded into population centers may have been more susceptible, but those living in towns and villages hosted their share, too. Bedbugs became an unavoidable part of life; tucking away in bedding, furniture, and walls. Full-grown bedbugs are about the size of a grain of rice and can squeeze their flat, grenade-shaped bodies into cracks, crevices, and seams, hiding by day and feasting by night. Their life cycle spans weeks or months. Once an egg hatches, the young nymph strikes out in search of blood, attracted by our scent, temperature, and other cues. ATP molecules (life's storehouse of chemical energy) in our blood encourage the bugs to engorge themselves.[7] To the bugs, our blood is like an addictive cocktail. As the insects grow, they shed their skin and increase in size with each molt. Over a period of months, a characteristic crumble of skins builds up in the corners of a nightstand drawer or along the seam of a mattress. By the time a bug reaches adulthood, it will have molted and fed upon a human host at least five times. If slumbering humans aren't available, the bugs can wait for weeks or months, surviving without breakfast, lunch, or dinner. If you have had the pleasure of ridding a child of lice, which feed daily or hourly, you know that bagging infested bedding for a week or two is enough to starve the pests to death. But the ability of bedbugs to persist without food makes them particularly difficult to control. With mature females laying several eggs a day, the number of active bugs can add up before we even know what bit us, making them a particularly insidious houseguest.

By the late nineteenth century, the United States Department of Agriculture urged vigilance, particularly for city dwellers; checking the "crevices and joints" of beds every few days and general cleanliness would help keep the bugs under control.[8] Clearing bugs from the home was part of the March tradition for one turn-of-the-century Arkansas farm family: "To slow the bedbugs down and thin them out, we took down and outside the beds and all the bedding, emptied the old straw ticks and burned the straw. We washed and boiled anything that was washable and scalded the bed slats and springs and poured boiling water in all the cracks and crevasses that the water would not ruin." Even after all the burning and boiling, and despite precautions like setting the bedposts in cans of coal oil and removing them from contact with the wall, any respite was short-lived. The bugs returned within weeks.[9] Most likely our grandparents or great-grandparents were well versed in keeping those "wallpaper flounders" at bay. So how and when did "spring cleaning" morph from bedbug control to simply clearing closets and dressers of outdated clothing?

"A Miraculous Insecticide"

Desperate for a good night's sleep, early-twentieth-century homeowners welcomed highly toxic products into their homes as chemical treatments became increasingly available. Mercuric chloride, benzene, sulfur fumes, cyanide gas, and even Zyklon B pellets (a form of cyanide subsequently used in Nazi gas chambers) promised some respite, even if this came at a risk to home and health. Then entomologists finally hit upon a chemical that worked *and* was relatively nontoxic to humans—DDT. It was nothing short of a miracle. Just after World War II, bedbugs virtually disappeared from the developed world. Although the chemical, a compound of chlorine, hydrogen, and carbon, was first synthesized in 1874, its insecticidal properties weren't discovered until 1939.

Seeking a cheap, effective agricultural pesticide, Swiss chemist Paul Müller rediscovered DDT only after testing hundreds of other chemicals. DDT was so lethal that flies dropped dead in their experimental cages shortly after contact with treated surfaces. The residues were so persistent that the cages had to be aired for a month before they could be used again.[10] In an age with few options for preventing or treating insect-borne diseases like typhus and malaria, DDT was a godsend. It targets the insect's nervous system by effectively propping open protein channels, allowing an endless flow of biochemical signals. Repeated and spontaneous firing of neurons ensues, followed by death. The chemical became the go-to treatment for lice, fleas, mosquitoes, and eventually bedbugs. Müller won the Nobel Prize.

In World War II military camps, typhus carried by lice was rampant, as was malaria, depending on the region. Controlling these insect vectors was critical. Not only could DDT do that, but the concentrations necessary to kill insects caused little observable toxicity to humans. Even better, DDT's persistence meant that it stuck around for weeks or months so the killing continued even as insects hatched out over time. Plus, it worked on contact rather than orally. This was particularly good for targeting pests that fed exclusively on blood. In these post–World War II years of industrial-chemical zealotry, some even suggested that with DDT humans may someday banish "all insect-borne disease from the earth."[11] By 1972, well over a billion pounds of the chemical had been applied to homes, gardens, wetlands, and millions of acres of US cropland.[12]

Of course, the story of DDT didn't end well. One of the chemical's most favorable characteristics, persistence, combined with its tendency to accumulate in the fatty tissues of insects, birds, fish, and mammals, made it the scourge of the twentieth century, as so eloquently revealed by Rachel Carson. In 1972, its registration for domestic use was canceled. Eventually, residues of DDT and its metabolites, detectable in both humans and wildlife for decades, subsided. Yet DDT continues to haunt us today in a most unexpected way.

Return of the Bedbug

For nearly five decades after the discovery of DDT (followed by other pesticides no longer sold here in the United States), we enjoyed relief from bedbugs. But those days are over. There isn't any one reason for the resurgence. Increased world travel and immigration, particularly from parts of Africa, Asia, and Europe where bedbugs were never really controlled; our own complacency; even demographics played a role. Because bedbugs breed stigma, infestation is a sensitive issue which can make dealing with it all the more difficult. "Reservoirs of bedbugs have been created especially in poor segments of populations," says Romero. "They cannot afford bedbug treatments, many live in low-income housing where resources for pest control is limited, many are undocumented and do not want to call attention with bedbug issues. These vulnerable segments of the population can leave bedbugs behind in public places where other people can get infested." The bugs are so omnipresent that in-home health-care workers are now trained to detect infestation and avoid tracking the bugs back to their own homes and offices.[13]

Even as many of us remained blissfully ignorant, bedbugs were never *fully* eradicated from the United States or Western Europe, and pockets remained throughout the 1980s and '90s. Romero says that though there were few cases, bedbugs were still around. How did they slip through the DDT era and beyond? During its heyday DDT had been applied to sheets, pillows, and bedding. By the 1950s, primarily because of DDT, bedbugs had become so scarce that researchers turned to other problems.[14] But in response to massive spraying, bedbugs, along with houseflies and mosquitos, did what bacteria, plants, and other animals have been doing since life began. They evolved—in this case, under selection pressure from a toxic chemical. And they evolved rapidly, aided by their short generation time and their formidable ability to go forth and multiply. Like penicillin, DDT was a short-lived miracle, with the first signs of resistance bubbling up in houseflies as early as 1946. But still the chemical continued to be used, and overused. A few years after DDT was marketed for bedbugs, they too evolved resistance. By 1957, resistant bedbugs could be found in locations around the world.[15]

Currently more than 570 insect species, from bedbugs to houseflies, mosquitos, and fleas, are known to be resistant to at least one insecticide; as with weeds and bacteria, by the time you read this chapter there will be more. Evolution has rendered some 338 different insecticides useless in one species or another. And, like antibiotics, chemotherapy, and herbicides, resistance to more than one insecticide is all too common.[16] The traits responsible for resistance are by now familiar. No matter the species, a limited number of options are available for surviving toxic threats: exclude, excrete, detoxify, sequester. While some bedbugs transform pesticides into harmless by-products, others acquire a mutation at the intended target. Enhanced capacity to metabolize and excrete DDT made the chemical ineffective. So too does a mutation in neuronal ion channels. And like other species, insects draw upon both standing genetic variation and novel mutations for resistance.

As DDT's efficacy declined, pyrethroids filled the void. These synthetic versions of pyrethrin, a natural insecticide produced by plants in the *Chrysanthemum* genus (often used for organic gardening), are now a common household insecticide. Like DDT, pyrethrin derails the conduction of nerve-cell signals; unlike DDT, the pesticide quickly degrades and so is short-lived.[17] Marketed in the 1800s as "Persian Insect Powder," pyrethrin offered a safer way to kill bedbugs. Pyreth*roids* are tweaked so that they stick around longer than their natural counterpart, enhancing their effectiveness. Today, there are over 3,000 commercial pyrethroid-containing products for home and garden. But bedbugs are no longer sensitive to the insecticide. In 2007, Alvaro Romero and colleagues designed a study to better understand the causes of the bedbug resurgence and resistance. "We initially thought that resistance to pyrethroids was created by overuse and inappropriate use of these products in the last 10 years," says Romero. The scientists found that of eight different bedbug populations collected from apartments across the country, most were resistant to pyrethroids. One population found thriving in Cincinnati, Ohio, could withstand more than *12,000* times the usual killing dose of the insecticide.[18] Of those eight "wild" populations tested, only one, collected in Los Angeles, California, was sensitive to pyrethroids; as were two long-time "captive" populations. One of these was derived from a population sustained on the blood of retired military entomologist Harold Harlan for more than forty years. Harlan's colony—famous among bedbug researchers—is one of a very few maintained throughout the bedbug-free decades. That resistance might have evolved in response to pyrethroids wouldn't be surprising. But that didn't appear to be the case.

As it turns out, ancestral resistance to DDT appears to have equipped today's bugs with resistance to pyrethroids as well; one mutation in the target site that confers resistance to two classes of pesticides. "We speculate that bedbugs could have maintained resistant genes for many decades," says Romero, reminding me that DDT is still used in some African and Asiatic countries, which may have aided retention. The most popular hypothesis for the resurgence of DDT-resistant bugs and for pyrethroid resistance, suggests Romero, is importation of resistant bugs from other countries where DDT is still used.

In addition to those target-site mutations, some resistant bugs ramp up defense by producing multiple copies of genes involved in chemical metabolism. Expressed in the bug's outer shell, the genes effectively create a futuristic pumped-up suit of detoxifying armor and block the chemical from even reaching its target—an elegant evolutionary solution.[19]

Biting Back?

Reluctant to fumigate her home with toxic pesticides, Abby took the advice of an herbalist friend and tried pennyroyal oil (a good example of how "natural" does not equate with "nontoxic"—in large enough doses, the oil is toxic to the liver and kidney). It didn't work. Despite Abby's penchant for a relatively low-impact, organic lifestyle, the bedbug situation required a more toxic solution. Her husband Andy called the exterminator. There are a few insecticide formulations with different active ingredients, including some botanicals that are labeled for use against bedbugs. Abby's house was treated with a combination of the old standby, pyrethroid, and a newer pesticide called Phantom (chlorfenapyr is the active ingredient). Rather than targeting the nervous system, Phantom starves the insect to death, even in times of plenty. The radically different mechanism provides it some protection from cross-resistance conferred by DDT. Still, it took four treatments, and cost nearly $1,500. "Each time the pesticide treatment didn't work," says Abby, "I felt like I was going to hell." Fortunately, the bugs that had invaded Abby's home had not yet evolved beyond the exterminator's reach. Phantom is now widely used for insect control. But, says Romero, the possibility of resistance certainly exists, and populations of insects other than bedbugs have already evolved resistance. So what are "we," the victims and exterminators alike, to do?

Just as we must reconsider how, when, or in what combination to use antibiotics, chemotherapy, or herbicides, we would do well to rethink insect control all together. Otherwise, we will find ourselves inundated not only with bedbugs but with lice, fleas, and agricultural pests. Just as is done on the farm field, integrated pest management, including nonchemical treatments, is working its way into our homes. Some approaches are modernized versions of ancient remedies. But rather than pouring boiling water onto our bed frames, an exterminator might seal up a room and raise the temperature. Bedbugs begin to die when temperatures reach around 115° F. Steaming or heating works well in small spaces or easily sealed rooms. But in Abby's airy Victorian home, heat treatment wasn't an option. Nor was the opposite solution: freezing bedbugs with a carbon dioxide "snow"—although that too may work in a smaller space. And then there are the bean leaves. Before toxic pesticides, Europeans staved off bedbugs by scattering bean leaves on the bedroom floor; the leaves' coating of dagger-like hairs skewered the bugs' legs. This elegant and nonchemical solution inspired researchers including entomologist Mike Potter and others to emulate it, creating a synthetic version of the hairy landscape. The polymer replicas snagged bugs, but only temporarily; yet with some improvement we may one day want to scatter some synthetic bean leaves about *our* rooms before turning in for the night.[20] Even so, though a physical approach might stymie the process of rapid evolution should bean leaves become all the rage, how long before a bean-leaf-resistant bedbug population arises?

Vigilance, too, provides some respite. Just as tick checks are part of the daily routine for those of us living in deer-tick-infested areas, if you travel a lot, you may want to lift the hotel sheets, examine the mattress, and check your suitcase before settling in at home. Perhaps we'll think twice before picking up that roadside couch—or, at the very least, give it the once-over before we do. "We don't let people bring their own futons into our house anymore," says Andy. "No more renters in our house; no used upholstered furniture; we avoid cheap hotels; and we assume that people who pick up used furniture on the side of the street are adopting bedbugs—this includes most of our friends. It's a bit emotional, but that is a central aspect of the 'BB experience.' They really mess with your head." While we won't be placing the legs of our bed in coal oil, there are now plastic bedbug "interceptors" that, at the very least, may help with bedbug detection; we might also declutter our bedrooms, vacuum the mattress, use an impermeable mattress cover, and seal cracks and crevices around the bed. The other day, I returned some bedding to the local department store and asked the cashier how they can be sure I wasn't returning it with a few hitchhikers attached. "We don't reshelve if the package has been opened," she said. "We send them back to headquarters." Lest we forget, we are all in this together. "Everybody is susceptible to bedbugs," says Romero, "even entomologists."

The story of bedbugs is little different than antibiotics, antivirals, herbicides, and any other human-versus-pathogen or human-versus-pest conflict, except in this case we *almost* had them beat—only to face a pest that has managed to regroup and return with better defenses. While the worst thing bedbugs may do is keep us up at night and drive us nuts, they are emblematic. Mosquitos, bollworms, ticks, beetles, mites, lice, and many other pests have all evolved pesticide resistance; some carry lethal disease while others cause billions of dollars in damage. Life is resilient. Yet even as those species that we wish to eradicate persist, too many others never slated for destruction—bees, damsel flies, frogs, and songbirds—face extermination, if not extinction. If our experience with pests like bedbugs teaches us one thing, it is that we do not have the power to pick and choose. We might provide the pressure, but nature does the selection.

PART II
Natural Selection in an Unnatural World

CHAPTER 6

Release: Toxics in the Wild

"When that *Science* article on tomcod was published," recalls geneticist Isaac Wirgin, "I'd never been so popular." His phone was off the hook with calls from the *New York Times*, *National Geographic*, National Public Radio, and even the Associated Press.[1] Atlantic tomcod, improbably celebrated by the Québécois each winter, are ugly little fish that also make their home in the PCB-laden Hudson River, fodder for the much more popular bluefish and bass. Wirgin and colleagues had just published a paper confirming that Hudson River tomcod had evolved resistance to incredibly high concentrations of polychlorinated biphenyls (PCBs), the chemicals responsible for turning the majestic Hudson into the largest Superfund site in the nation. Over a period of 30 years, General Electric, the company that promised "We Bring Good Things to Life," released an estimated 1.3 million pounds of PCBs into the Hudson.

As with DDT, one of the more useful qualities of these chemicals became one of their most infamous characteristics—persistence. PCBs collect in the fat stores of crustaceans, fish, mink, and fish hawks that make their living along the Hudson, accumulating as it works its way up the food web. Tomcod accumulate screamingly high concentrations of the chemicals. When Wirgin's group first reported their findings, Wirgin became a Scientific Rock Star for the day. Not only did they show that the fish evolved rapidly but, along with others, they traced chemical resistance to a change in just two amino acids—small changes in a receptor normally involved in growth, development, and reproduction but one that could also combine with PCBs. This was evolution in action and it had happened not in bacteria or insects or weedy species—but in fish, a vertebrate species more closely related to us than any of those rapidly evolving pests and pathogens that filled the pages of earlier chapters in this book, reminding us that rapid evolution isn't just for the spineless.

Wirgin's decades of research brought pollutant-driven evolution into the public spotlight. But it was not the first discovery of such an event. Like pesticides and antibiotics, pollution is a powerful selective pressure—altering the genomes of untold numbers of species in contaminated rivers, coastal regions, and even on land. Back in 1994, geneticist John Bickham and biologist Michael Smolen raised the specter of evolution in response to pollutants caused by increased mutation or by selection, predicting that "most toxic chemicals in the environment will affect evolutionary processes."[2] Yet our society has an addiction to the products of industrial chemistry and, try as we might, realistically there is no turning back. As we force nature's hand there will be winners and losers. We might call those that adapt, whether weeds, pathogens, or tomcod, evolution's winners. But we also know that winners, like steroid-addled athletes, are not always what they seem to be. And too often the fallout of a chemically altered life reaches well beyond the individual. This is the story of those so-called winners—species that, like the tomcod, have evolved to resist chemical pollutants.

Industrial Evolution

Claude Boyd was just a masters student at Mississippi State in 1962 when he and his advisor, Denzel Ferguson, made their first trip to a pesticide-soaked cotton field in search of survivors. That year Rachel Carson's *Silent Spring* was changing the way the nation viewed its relationship with pesticides. At the time, recalls Boyd, "there was a culture of using a lot of pesticides. Everyone knew they were probably using too much, but nobody wanted to stop."[3] Among the many arguments Carson had made about the misuse of pesticides was the rampant evolution of resistance in pest species. "To the list of about a dozen agricultural insects showing resistance to the inorganic chemicals of an earlier era," wrote Carson, "there is now added a host of others resistant to DDT, BHC, lindane, toxaphene . . ." and the list went on.[4] "Darwin himself," wrote the biologist, "could scarcely have found a better example of the operation of natural selection."[5]

Denzel Ferguson, says Boyd, figured that if pest species could evolve, why not other species? In those days, DDT was everywhere. "I'd go to bed—this was before there was air conditioning everywhere—and everybody would shut up the windows and spray the house with DDT, go outside for a while, and then go back in. It was a common practice." Even the radio station WDDT in Greenville, Mississippi, seemed to celebrate the pesticide. "Back then," says Boyd, "the towns had mosquito foggers; and I guess they had DDT in those and they just went down the streets every afternoon. They just sprayed it everywhere." In particular, the farm fields along the Mississippi Delta were treated repeatedly during the growing season with a list of pesticides that today reads like a who's who of banned chemicals. Spraying continued even as the chemicals began failing on common crop pests like the boll weevil. Resistance be damned, farmers then as now fought back with increasing concentrations and next-generation pesticides. The effects were devastating, as recounted in one of the lab's first publications, authored by Boyd, Ferguson, and graduate student S. Bradleigh Vinson. "Although no attempt was made to make a complete count of dead animals, the following were noted: large numbers of southern cricket frogs, several green tree frogs, a fowler's toad . . ." and the list goes on.[6] Fish, frogs, turtles, snakes, butterflies, beetles, snails, and birds had succumbed to DDT and other pesticides. Yet amid the mass mortality were signs of life. That was in 1963.

At the time, there were hints here and there that resistance wasn't just for weeds, worms, and bugs. In 1960, the journal *Nature* published a short note about a rat population on a farm in Scotland apparently "more than normally resistant" to rat poison;[7] and in 1962, scientists artificially selected resistant laboratory mice by exposing them to DDT and breeding the survivors. But rapid evolution in wild populations that had the misfortune to make their homes on increasingly contaminated farm fields, ponds, lakes, or suburban backyards was unheard of. Ferguson, says Boyd of his advisor, was a young scientist determined to distinguish himself. And those animals living in the farm fields of the Mississippi Delta were virtually steeped in pesticides.

If anything other than insects was going to evolve rapidly, those fields, ponds, and lakes would provide the perfect opportunity. As the scientists observed, there were plenty of dead and dying animals. That was to be expected, as the fields, after all, were saturated with poison. But among the dead was the flash of small fish and the sound of cricket frogs. Subsequent laboratory studies confirmed that frogs had indeed acquired resistance to DDT and that fish had acquired resistance not only to DDT but to several other pesticides. Research also suggested that resistance passed on to the next generation—a hallmark of evolution. In resident mosquito fish, even the second and third generations were resistant.[8] "If genes and physiological mechanisms for resistance exist in vertebrate populations," they wrote, ". . . sufficient selective forces are present . . . to increase their frequency." Subsequent studies revealed resistance in populations of sunfish, shiners, and bluegills. If the ponds had been treated for 10–15 years, as they had by 1960, for species requiring a year to mature, this meant that the population had evolved in just ten generations or so. That DDT-resistant laboratory mice evolved tolerance after just seven generations provided further evidence that some vertebrates could evolve incredibly rapidly and could do so in response to toxic chemicals.[9] Boyd and Ferguson suspected that rapid evolution was likely aided by the existence of genes already present in the population that contributed to detoxification, or excretion, or that were otherwise able to protect against these novel chemical threats. Theirs was the first report in the scientific literature confirming that vertebrates had evolved rapidly in the wild in response to a chemical designed to kill. This was big news.

Half a century later, Boyd recalls that just getting the paper published was an "eye-opening introduction to the real world." One journal rejected the work outright as being "contrary to the editorial policy to publish anything *good* about pesticides," while another accepted the paper, apparently putting it on a fast-track for publication. "As far as we could see," recalls Boyd, "it wasn't even reviewed—because they assumed it *was* something good for pesticides. Who is to say that resistance to pesticides in a fish is a good or bad thing?" In the end, the narrow-mindedness on both sides of the issue turned Boyd away from a career in environmental toxicology and toward aquaculture (where he has had a long and distinguished career). That was 50 years ago, yet the question—is resistance good or bad?—remains relevant and is perhaps even more so today, amid mounting evidence of wildlife species resistant to industrial-age chemicals and other human influences. If vertebrates can evolve their way around chemical toxicants, do we need to worry so much about contamination? One evolutionary biologist I know commented: "I think it's sort of hopeful. Populations will survive, no matter what." But if populations can evolve resistance, why spend millions on cleanup? Is there a downside to pollution-driven evolution?

Bringing Better Things to Life?

Industrial pollutants—metals, PCBs, dioxins, combustion products, and who knows what else—can all select for resistant populations. Around the time Wirgin started his work with tomcod, there were early reports that a common coastal minnow, the killifish, had acquired resistance to dioxin (far more toxic but structurally similar to some PCBs) and to methyl mercury, but heritability of resistance had yet to be demonstrated. Those earlier studies certainly hinted at evolution—but without the heritability card, they remained suggestive rather than conclusive. That killifish were resistant to a chemically altered environment really was no surprise. The species seems almost indestructible. Tolerating the ebb and flow of tides, killifish survive in shallow puddles baking in the summer sun, and they thrive in both high- and low-salt environments. Even so, resistance to sun, salt, and other natural challenges need not confer resistance to toxic chemicals. Yet, following a series of elegant and painstaking studies, the aquatic toxicologist Diane Nacci and colleagues including Andrew Whitehead demonstrated that the minnows also resisted PCBs. Not only that, but distinct populations of killifish inhabiting PCB- and dioxin-contaminated creeks and bays all along the northeastern coastline had evolved resistance. This was rapid evolution over and over and over again. And, as Nacci, Whitehead, and others point out, there is something profoundly different about evolution in response to PCBs and dioxins. "Insecticides," they write, "were specifically designed to poison and kill target organisms by interfering with specific molecular targets, whereas toxicity to wild fish from dioxin pollution was not intentional."[10]

It is becoming increasingly apparent that living things have a deep well of genetic variability to draw upon when they are exposed to chemicals, particularly those that, like some pesticides and many antibiotics, are based on naturally occurring structures. But this is evolution in response to a novel synthetic chemical and it is happening in the wild, not in a farm pond. This is evolution unbounded. So it seems even more bizarre that the pathway to such rapid evolution likely rests upon the selection of ancestral genes lurking in the population rather than some new random mutation. Yet in a survey of tomcod from the Hudson River and along the eastern coast of North America, Wirgin found resistant gene variants in a small number of fish from relatively unexposed populations near the Hudson. "I think that if you went back historically prior to deposition of chemicals," says Wirgin, "you probably would have seen the same in the Hudson." This means that the genes for resistance already existed but that they increased in frequency under the selection pressure imposed by PCBs. This standing genetic variation is believed to underlie many cases of rapid evolution, including pesticide resistance in mosquito fish. And the phenomenon provides a sliver of hope for populations under threat. That is, if they can manage to survive long enough for this sort of genetic rescue.

The Hudson River has been contaminated for well over 50 years with PCBs and other toxic chemicals. The best guess, for now, is that resistance traits evolved within 50 years. That is exceedingly fast for species that breed annually. But could it have happened even more quickly? Within a decade or two, as it did for Boyd's frogs and mosquito fish? Or even a year? One of the most comprehensive and convincing studies demonstrating rapid evolution in vertebrates comes not from contaminated farm fields or rivers but, fittingly, from Darwin's natural "laboratory"—the Galapagos Islands. For 30-plus years, evolutionary biologists Peter and Rosemary Grant have recorded beak size, body shape, and environmental conditions in a population of Darwin's finches. Over those three decades the couple witnessed the dynamics of evolution, catching it in action, as beak and body size shifted between generations under naturally occurring environmental fluctuations. Their work is a testament to rapid evolution in vertebrates.[11] The Grants revealed evolution in a species far less productive than rats, mice, or fish, writing that "natural selection occurred frequently in our study, occasionally strongly in one species and oscillating in direction in the other. . . ."[12] Most likely, standing genetic variation there, too, helped speed things along.

That ancestral genes can rescue a population faced with randomly recurring environmental change makes sense. Genes that come in handy are selected every once in a while over and over again, and so maintained in the population. It may even make sense that some ancestral genes may be responsive to pesticides, particularly for pesticides derived from naturally occurring toxicants. But how and why ancestral genes aid populations exposed to novel synthetic chemical pollutants like PCBs or dioxins is anyone's guess.

So too is how *quickly* a population can evolve, particularly as we alter the earth's environments at an increasingly faster clip. The Grants observed changes within an astounding single generation. How quickly can a population respond to PCBs or dioxins or other toxic chemicals? While no evolutionary biologists had the foresight to set up shop along the banks of the Hudson as PCBs began making their way into the river's sediments and water, existing historical fish collections may hold the answer. Since the 1930s, Hudson River tomcod have been captured, fixed, and bottled up, stowed away in the archives and shelves of natural history museums around New York. The answer is there for the taking, and Wirgin, like a kid with his nose pressed to the glass, wants in. But, says Wirgin, the formaldehyde used for preservation is the bane of DNA analysis. For now, those pickled tomcod are holding tightly to their history and will continue to do so, at least until the right technology comes along.

Millions of Miles to Travel

I am compelled to share one other example of contemporary adaptation in response to pollution, because it reminds us that chemicals do not only come from far-off factories. A toxic brew of salts, metals, and oil is generated by a simple act most of us do every day: driving. In the United States there are 4 million miles of road; some 160,000 miles of asphalt highway; and on average, Americans drive some 13,000 miles a year.[13]

Roads cover roughly 1 percent of the nation's land. Laid down over the past 50 or 60 years, those millions of miles of paved roads created a complex network for the distribution of chemical contaminants, from city to suburb to countryside. When the winter snow starts to fly, so too does the road salt as trucks trundling down quiet roads and busy highways alike may dump some 30 tons of road salt *per lane mile* (one 12-foot lane, for one mile), annually.[14] Each year the spring rain and snow melt carries those salts, along with toxic metals and hydrocarbons released from the tailpipes of our cars, rubbed from our brake pads, or worn away where rubber tires hit the road, into the surrounding environs.

So it is no surprise that even in the quiet hemlock and oak forests of northeastern Connecticut, evolutionary biologist Steve Brady finds amphibian populations compromised by roadside runoff. "They are occupying these ponds that are receptacles for all the stuff coming off the road," says Brady, clutching a handful of glistening spotted salamander eggs collected from one of those roadside ponds.[15] "The chance of survival here is much lower than that in a woodland pool. A little over half the eggs survive the first 10 weeks of development." Those first weeks are a major hurdle, especially for wetland amphibians that already face intense natural challenges. Under the best of conditions, as spring progresses and the pools naturally dry up, young salamanders that haven't yet sprouted legs may die. As roadside pools dry up, contaminants become increasingly concentrated. How did salamander populations, which tend to stick close to home, survive year after year in such a toxic environment? Lacking any specific gene or receptor to target with genetic analysis and unable to raise successive generations in the laboratory—at least within his lifetime as a graduate student—Brady carried out an elegant sort of study: a *Switched at Birth* for salamanders. He transplanted subsamples of eggs from one environment (either roadside or woodland) to the other. Then he watched how the creatures fared through hatching and their early stages of development.

In the end, the roadside salamanders out-survived woodland salamanders in a pattern that suggested the roadside populations had become locally adapted to their harsh conditions. Like those killifish, salamanders survive in pools that come and go with the season and naturally become concentrated as the weather warms up. So it may not be all that surprising that a population may possess the genetic wardrobe to cope with increasing salt levels—even when the salt enters the environment by unnatural means.

Brady's finding adds another species to the list of evolvable vertebrates. The studies are stacking up, and as they do, they are burying the old notions of evolution that many of us carry with us—that slow march through the strata of time. Life's genomes are more fluid than once believed. We notice this especially when genetic variants, like ghosts from environmental challenges past, offer populations the flexibility to withstand rapid environmental change. And yet, rapid evolution is relative. Selection acts on genes and traits passed from one generation to the next. A species that requires 10 years or so to mature, like beluga whales (of which some populations are now loaded with pollutants), would be hard-pressed to make their peace with the rapid onset of toxic chemicals in their environment. And even species with more-amenable life histories, like those that spawn annually and produce hundreds of eggs, don't necessarily adapt. Consider the wood frogs that live alongside the salamanders in Brady's woodland and roadside ponds. Unlike those of salamanders, says Brady, roadside frog eggs are even more compromised than woodland frogs when placed in contaminated ponds, and they don't do so well in woodland ponds, either.

So why didn't those sickly populations go extinct? Brady suspects that the roadside populations may be sustained by healthier woodland frogs filled with wanderlust, traveling from their cleaner woodland ponds to roadside ponds, which leads them to set up home under less than favorable roadside conditions. Not exactly a satisfactory explanation, says Brady. The population is so stressed that it *ought* to go extinct. And there's another thing. The frogs curiously take on excess fluid. "Roadside animals are edemic," says Brady, "like someone took a basketball pump and pumped them up. I've never seen this in a woodland pond. I see it driving around, too. I pull over and see what's out there, and sure enough, there's a big inflated wood frog."[16] Perhaps, says Brady, it is another consequence of roadside salts. Yet if frogs are migrating from cleaner woodland ponds to contaminated roadside ponds, it is difficult to imagine how they become compromised so quickly. "Roadside ponds could be like the Island of Misfit Toys, collecting all these misfits that don't do well in the other habitats because of competition or predation." So they make their way to the less desirable and highly toxic roadside pond.

The one thing that the story of frog and salamander makes clear is that there is no easy answer. When faced with the pressures of toxic chemicals, different populations and different species respond in different ways. Rapid evolution, says Brady, "is this ingredient in the recipe of life that influences the traits, the performance, the growth, the development, the shape, and the size of all the organisms that we are concerned about." We are far from predicting who will and who won't adapt. Species or populations with the right genetic baggage, like the tomcod or killifish, will adapt. But if they must wait for a new, randomly generated trait to confer resistance, the chances of it popping up and spreading through the population are slim—particularly in small populations that live long but don't necessarily prosper. And even in species that do play fast and loose with their genes, sending hundreds, thousands, or even millions of offspring into the world each season, there may be constraints to evolution. If the population is relatively inbred, perhaps caught behind a dam or isolated by a new Interstate off-ramp, it may simply be lacking in genetic variation. Or perhaps the dominant pollutant targets a critical enzyme that cannot afford to swap out one or two amino acids, as Wirgin's tomcod does. Or a single gene may encode a protein that has many functions, so that while tweaking it may avoid chemical toxicity, its other functions may be lost. Or a population might experience intermittent exposure to toxicants—perhaps as they migrate to and from a toxic site. All these possibilities raise a few nagging questions about rapid adaptation. Is it all good? Or are there costs when populations evolve, particularly in response to chemicals that are broadly toxic? And who pays? What about species that remain sensitive yet are migrating upriver, or feeding at the river's edge?

The Cost of Survival?

That there are costs to resistance, says evolutionary toxicologist Paul Klerks, is "pretty much a dogma, based on literature on pesticide resistance."[17] As with cash-strapped cities and towns, there is very little "fat" or excess energy in life's budget. Should there be any excess energy, it is put into reproduction and development—optimizing fitness. When a population must adapt to sudden change like metals or pesticides, depending on the mode of adaptation, energy may be diverted from reproduction to defense. Perhaps additional copies of a protein that pumps toxic chemicals from the cell or enzymes that chew up a pesticide are produced; or a receptor is altered just enough to prevent a chemical from damaging cells but becomes less efficient when carrying out its normal activity. In these cases the next generation may be left footing the bill in the form of reduced numbers, slower growth, or less fitness in general. Sometimes a change for the better in one scenario is a change for the worse in another—perhaps offspring are more sensitive to sunlight, or less tolerant of other everyday kinds of stress. And sometimes the cost comes as a loss of genetic variation. As resistant traits are selected, others will likely be lost. In extreme cases, a population that loses genetic variation could end up like inbred European royals.[18] Or like the Florida panther, inbred and unhealthy until "rescued" by an infusion of genetic variation. "But," adds Klerks, "I believe that it is also well accepted that there are cases where costs are not obvious, and these may be seen more in the recent literature compared to the past."[19] And, suggests Klerks, even if there is a cost, natural selection may once again kick into action, selecting for resistant variants that essentially cost less.

Environmental conditions are constantly shifting, both as a matter of course and as a result of our activities. So as killifish or tomcod tolerate PCBs and dioxins, how will they tolerate temperature shifts? What of the many other chemical pollutants that are present? The examples I have used are dominated by one major stress, but most environments are contaminated with complex chemical mixtures or other stressors. While there are some examples of multiple resistances—a protein that confers resistance to both metals and heat stress, for example—the more likely scenario is that the genome is pulled in many different directions by all manner of chemical and other stressors. Even as we are finding that some enzymes and proteins can carry out multiple duties, if selection demands attention to one over another, subtle imbalances may emerge—like a good utility soccer player who covers goal but is then out of position to score one for the team. The selection of a trait conferring resistance may also drag along other linked traits—genetic "hitchhikers"—fixing one problem while possibly creating others.

One intriguing tale of survival and cost comes from the yellow perch inhabiting lakes contaminated by decades of copper smelting in the Rouyn-Noranda region of Quebec. These are fish that have adapted to live with the toxic metals but are paying with their lives. Rather than spending precious and scarce energy on detoxification, these populations put their energy into ensuring there is a next generation. They mature and spawn early and have shortened life spans. They are, in other words, living fast and dying young.[20] Even when a population manages their metals while enjoying a normal life span, there may be hidden costs. One metal-resistant population of killifish, evolved under laboratory conditions through a process of exposure and selection, can survive conditions that certainly would have killed their ancestors. But when they are placed in a cleaner environment their ability to be fruitful and multiply is compromised.[21] They are also more sensitive to increased temperatures—these days a more likely scenario than a suddenly clean environment. Whether such laboratory findings are transferable to the wild isn't yet known, but they may well apply to the brown trout living in the River Hayle, in the southwest corner of England. Like other extraordinarily resistant species, these fish withstand concentrations of metals that would have been lethal to their preindustrial ancestors.[22] And they appear to do so by dipping into the detox trough, relying on pumped-up detoxification and tolerance—a reallocation of energy. Whether or not these kinds of costs may become a burden, should the weather warm up a bit, remains to be seen.

Meanwhile, Nacci and colleagues find few costs in their populations. "Bear in mind," says Nacci of the loss of receptor sensitivity, "that this is a 'null' mechanism, whereby function is poor or lost." There isn't a cost in terms of churning out multiple copies of genes and associated proteins or diverting resources to pump out noxious chemicals. But what about the "normal" services provided by that less sensitive receptor? Perhaps the change occurs in a way that enables it to remain responsive to natural signals. Or not. And then there is the loss of genetic diversity—a bad thing no matter how you slice it, especially as we begin to see how important standing genetic variation is for adaptation. So far, that hasn't played out in any obvious way in Nacci's resistant populations—yet.

The evolution of resistance in the wild is a mixed bag. Some species have the genetic goods to withstand the challenges of both nature and humans; while others, more set in their ways, fare worse when challenged with rapid change. And in ecosystems shared by both fast and slow (or non-) adaptors, predators that are slow to adapt yet dependent upon "fat and happy" resistant prey for their meals may face an additional challenge—a parting shot of sorts from prey to predator.

The Selfish Trait?

Years ago, Denzel Ferguson wondered about some of those other species inhabiting contaminated farm ponds—the ones that might not adapt so quickly and that may feed upon those who do. So he fed pesticide-loaded and still-living mosquito fish to redfin pickerels, bullfrogs, turtles, cottonmouth snakes, and other potential residents. The diminutive mosquito fish had accumulated enough pesticide to kill its predators. The fish had become so toxic that Ferguson feared for the capricious child who might play at fishing and then make a meal out them.[23] PCB-resistant tomcod pose a somewhat similar problem. "This is not a rare species," says Wirgin. "You've got these little tomcod swimming around that are souped up with high levels of PCBs and dioxins and probably other things." The tomcod have become a step in a biological conveyer belt, transporting and concentrating PCBs and other contaminants from water and sediments, up the food web and into the bodies of eels, hake, bluefish, striped bass, and others. Yet despite decades of research, we know little of how those species tolerate PCBs acquired in this way. If the contaminants don't cause outright reproductive, developmental, or neurological toxicity, carrying around a load of PCBs certainly adds a biochemical stress. For species already stressed by commercial and private fishers and an increasingly unstable climate, the combination just might send them over the edge.

Contaminated tomcod or other similarly contaminated fodder fish are not just a problem for other fish. Snapping turtles, kingfishers, and American mink may all be on the receiving end of a potentially toxic meal brought up from the sediments and water by way of more-tolerant species.[24] Mink are exquisitely sensitive to PCBs, and one effect is reproductive failure. In one study, mothers fed any more than the smallest amount of Hudson River fish (and thus the lowest PCB dose) lost all of their kits by ten weeks.[25] Forty years after the ban of PCBs and billions of dollars of cleanup later, for mink and most likely other slow adaptors, the majestic Hudson River remains a tough place to raise a family. While tomcod enjoy the benefits of resistance, it may be that other species, like mink, are picking up the tab. "I wish I could say there was a cost [to tomcod] at the population level," says Wirgin. "My guess is that there is no cost. But at the community level, the cost is huge." Until PCBs are no longer accessible to tomcod and their prey, the chemicals will continue working their way up the food web and into more-sensitive species.

Sometimes resistance can have odd, unexpected effects. Piles Creek, New Jersey, is a mess of PCBs, hydrocarbons, and mercury. For the past 20 years or so, marine biologist Judith Weis has studied tolerance to highly toxic methyl mercury and the interactions among killifish, grass shrimp, and blue crabs—denizens of this industrial-age creek. In killifish, methyl mercury tolerance manifests as improved embryo survival. But once resistant embryos hatch out, growing into larvae and adults, they are as sensitive to the toxic metal as killifish from cleaner locales. Populations are sustained, but the adults are a bit off. "We stumbled upon the fact that the adult killifish from Piles Creek did not capture prey or evade predators as effectively as fish from clean places."[26] Such altered behaviors, Weis believes, contribute to reduced growth and shortened life spans for the adults. Within that system, grass shrimp are a typical prey item for killifish. Grass shrimp, like killifish, are mercury-tolerant, but apparently they lack mercury-induced behavioral deficits. "So," says Weis of the shrimp, "their major predator is a lousy predator, and they are still good at avoiding predation and are larger and more numerous—because they live a long, happy life and reproduce more." Meanwhile the blue crabs, also resistant, show a similar pattern as the killifish—they are mercury-tolerant but less-than-stellar predators. "They eat a bizarre diet largely of detritus, mud, and algae. They are supposed to be predators!"

Piles Creek offers up a disturbing future scenario. A system that upon first glance seems normal, perhaps even comforting, where life manages even in a grossly polluted site. Yet it is anything but normal. Instead, it has become a topsy-turvy world full of chemically addled survivors. Stricter environmental regulations, at least since the 1970s, have certainly helped to turn the Hudson, Piles Creek, and other sites into less-toxic relics of our careless past. But we are not off the hook—far from it. As long as we continue producing and releasing tons of toxic chemicals including mercury, pesticides, plastics, and antibacterials into Earth's environs, they will continue settling out into creeks, rivers, lakes, and landscapes. And life—some life—will continue adapting.

Survivor: Planet Earth!

Earth has always been a toxic place—bombarded by the sun's ultraviolet rays; flooded with highly reactive oxygen; infused with toxic metals like mercury, zinc, and cadmium; not to mention myriad poisons produced by life itself. Over billions of years, life evolved the means to sequester, excrete, detoxify, and in some cases even profit from many (but certainly not all) of these naturally occurring toxic chemicals. Yet in a blink of evolutionary time humans have reshaped the landscape, rearranged major food webs, and altered both the climate and the chemical environment. In the attempt to "live better though chemistry" we have created, produced, and released hundreds of millions of tons of tens of thousands of chemicals into the earth's environment. Many are no different than nature's own, while others are new to life. We are responsible for the release of a larger amount of a greater variety of chemicals in a *shorter period of time* than perhaps has ever before occurred in life's long history.

In their 2011 article entitled "Has the Earth's Sixth Mass Extinction Already Arrived?" Anthony Barnosky and colleagues write that our modern-day activities provide "more extreme ecological stressors than most living species have previously experienced."[27] Ever. Chemical pollutants, combined with altered landscapes, climate change, invasive species, and overfishing, have brought us to the brink of the earth's sixth mass extinction. Some say we are already there.[28] Currently, nearly 21,000 species around the globe are at risk of extinction. Among the most susceptible are large-bodied species, as well as the less fecund. While toxic chemicals may have a smaller impact than climate change, overharvesting, or dramatically altered landscapes, they are far from irrelevant. We produce some 500 billion pounds of plastics a year, and by some estimates a few hundred billion pounds of plastics now circulate throughout the earth's oceans. Roughly 1–12 percent of the billions of tons of PCBs released into rivers, streams, and oceans still cycle the globe more than 30 years after being banned. We add 2,000 tons of mercury to the atmosphere annually. And we release billions of pounds of pesticides worldwide.

If we simply keep on keeping on, one thing is certain: life will respond and the outcome may not be to our liking. Some species will go extinct and others will survive. Some will evolve and adapt to live in a severely altered world. It doesn't have to be this way. In the words of Barnosky and his colleagues, "The huge difference between where we are now, and where we could easily be within a few generations, reveals the urgency of relieving the pressures that are pushing today's species towards extinction." We would also do well to relieve the pressures that are pushing today's species to evolve. We can do better. If we need yet another reason to reduce our chemical footprint—this is it.

CHAPTER 7

Evolution: It's Humanly Possible

On a cold Wednesday evening after a presentation on pollution and evolution at Boston's Arnold Arboretum, I was asked a question I've struggled with ever since I began thinking about toxicants and evolution. I'd just rattled off a list of synthetic chemicals that are now part of our chemical environment: plastics, pesticides, flame retardants, PCBs. Usually after a presentation, the discussion revolves around how we can do better: improve toxicity testing, regulation, and management, or rely on green chemistry to create less harmful chemicals. But this time, perhaps making a bit of mischief, an audience member asked, "If we can't rein in these chemicals, why bother? Why *not* let nature take its course? You know, survival of the fittest?"

I hemmed and hawed, trying to come up with an intelligent answer. We have, of course, made great progress cleaning up our act over the past several decades. Since *Silent Spring*, we have tackled the obvious: rivers no longer catch fire and the more egregious chemicals no longer flow from industry pipes and stacks or contaminate our food and drink. But still, we produce and release plenty—from pesticides to antibiotics to the chemicals permeating household products. We live in an age of industrial chemicals and are unlikely to kick the habit anytime soon. And we know very little about the long-term effects of human exposures to small amounts of many different kinds of chemicals. I wanted to say, *Well, obviously we should do better, just because no one wants to live in a world where we eat and breathe toxic chemicals.* But the fact is we are already there. As Bill McKibben writes, this "Eaarth" is not the same Earth that was our home planet.[1] Dozens of industrial- and consumer-use chemicals already run through our veins, settle out in our brains, our fatty tissues, and deep within our bones. They have been associated with diabetes, obesity, learning disorders, and infertility. And evidence is growing that these chemicals will affect not only people today but the next generation tomorrow, and in some cases the generation after that. We are just beginning to discover the long-range health consequences of synthetic chemicals. But is it possible that we—like the bacteria, weeds, fish, and other species discussed in previous chapters—could adapt to these pollutants? Are we still evolving, and if we are, is it possible we too might grow less sensitive to our toxic surroundings? Will we be among the fittest?

Subjects of Evolution?

If you want to get evolutionists talking, just mention human evolution. Are we or aren't we still evolving? In the fall of 2013, David Attenborough—who for much of the 1980s, and beyond, was the voice of nature and evolution, bringing television programs like *Life on Earth* and *The Living Planet* into living rooms around the world—offered this opinion: "We stopped natural selection as soon as we started being able to rear 95–99 percent of our babies that are born. We are the only species to have put a halt to natural selection, of its own free will, as it were." His words circulated the globe in a flash. Several news outlets carried Attenborough's message forward: *Humans are no longer evolving!* But within days his words became blog fodder: "Attenborough's muddled thinking can't stop human evolution"; "Attenborough is wrong"; "Sorry Attenborough, humans still evolve by natural selection." Even *Science*, one of the premier academic journals got into the act, as Attenborough's words were highlighted and filed under "They Said It."[2]

Yet Attenborough, in suggesting that we are as good (or as bad) as we're ever going to get, is not alone.[3] Others, like anthropologist Ian Tattersall and geneticist Steven Jones, have also declared an end to human evolution.[4] Tattersall told Ira Flatow during an interview for National Public Radio that our large population has "simply too much genetic inertia," so unless there are major demographic changes, the human species is unlikely to undergo change.[5] Likewise, Jones argues that larger populations, increased mixing, and changes in reproductive patterns and how we live today mean that "all three parts of the Darwinian machine—mutation, natural selection, and genetic drift—have lost their power."[6] And it's not just a couple of outspoken academics who think we're going nowhere. In a commentary about human evolution written almost a year before Attenborough's declaration of independence, behavioral ecologist Denis Réale recalled the response of high-school students *and* university professors when asked if they thought humans are still evolving. Some 80 percent of the audience said "no."[7] It is true that we no longer need to grow thick fur to colonize the Antarctic; we build amusement parks where yellow fever and malaria once reigned; and those of us who wear glasses manage just fine, thank you, as we pass along our defective vision—much to the dismay of our teenage offspring. We manage with science and technology, removing the pressure to select for genes encoding thick fur or good vision. So it is understandable that some might think we've become untethered from nature—and from one of the driving forces of evolution, natural selection. But Réale and colleagues including population geneticist Emmanuel Milot, who have recently published on human evolution, don't see it this way.

Despite disease and environmental hardship, most of us in the industrialized world not only survive but thrive. But that isn't the case for all and certainly not for populations where disease still strikes down the young. "As long as some people die before reproducing or reaching reproductive age," says geneticist Chris Tyler-Smith quoted in a *Science* commentary, "selection is likely to be acting."[8] Not only is natural selection still ongoing but selection pressures change as environmental conditions shift around us. Emmanuel Milot says that in Western societies, while selection through survival is less important, if you change the *selective environment* (a thing we tend to be a little too good at) "different responses can be expected."[9] Essentially, says Milot, culture and technology, rather than *inhibiting* change in human populations, may instead *contribute* to the evolutionary process. Given new backdrops, or as Milot and colleague Fannie Pelletier write, "new playgrounds," from synthetic chemicals to climate change, natural selection can still act. "The traits playing the game," write the pair, "could be as diverse as cholesterol levels, age at reproduction, body shape, personality, immune defense, or even political choices."[10] If Milot and colleagues are correct, human populations haven't reached an evolutionary dead end but instead are creating conditions that, for better or for worse, may contribute to our evolutionary journey.

Not So Fast

How we live and what we do *can* have profound influences on selection, yet many of the very noticeable shifts in human traits happen not over a few generations, but rather several dozen or more. A shift to city living, for instance, may have contributed to the emergence of populations resistant to urban diseases like tuberculosis and leprosy.[11] And before that, as our ancestors hunted and gathered their way out of Africa, adapting to an agrarian lifestyle, some populations acquired the ability to drink milk well into adulthood. The evolution of lactose tolerance is a story that reveals not only how a cultural change—dairying—influenced human evolution but also how human population growth may have provided ample opportunity both for mutation and for selection to act upon all that new variation.

When our Middle Eastern or European ancestors first began milking livestock—whether cows, goats, or sheep—they were not chugging down fresh milk but rather fermenting the fatty, protein-packed liquid into yogurt and cheese products.[12] In most mammals, including many humans, tolerance for the milk sugar lactose ends with weaning, when the enzyme lactase is essentially turned off. But milk is a good source of fat and protein and, unlike meat, it is effectively a renewable resource. Biological benefits were to be had, and the few who happened to possess a gene variant for a form of the lactase enzyme that didn't quit went forth and multiplied, spreading their genes throughout the population. The kickoff event for this evolutionary shift, dairying, was cultural. The expanding population, which carried a greater load of rare genetic mutations—ripe for selection—likely helped. The beneficial effect notwithstanding, though, the process probably played out over thousands of years and many generations.

The interplay between malaria and the human genome is another case of a relatively recent adaptation spurred by cultural change. Malaria is believed to have become particularly important in early agrarian populations. As human populations settled down some 10,000 or so years ago, local landscapes became amenable to malaria-carrying mosquitos, which in turn bit nearby humans and infected them with malaria. Debilitating and sometimes fatal, malaria attacks healthy red blood cells, imposing intense selective pressures on exposed populations. The increased disease pressure selected for a number of genetic adaptations in humans. Most notable is the sickle-cell trait. The result of a single mutation, blood cells with the sickle shape provide resistance to the malaria parasite. But the cost for those with two copies of the mutant gene is sickle cell anemia, a potentially devastating condition. Because individuals carrying only a single copy show few adverse symptoms yet are still protected, the sickle cell gene has persisted in susceptible populations.

One other notable example of relatively rapid human evolution is an adaptation that enables some populations to breath easily at altitudes that leave the rest of us sucking air. The Tibetan plateau, 14,000 feet above sea level, where oxygen concentrations are some 43 percent of sea-level concentrations, is a harsh place to live. If I were to travel to Tibet, my red blood cell count, along with hemoglobin—the oxygen-ferrying iron component of red cells—would increase as I acclimated. This would compensate for the lower oxygen but would also create a sort of "rush hour" for red cells, thickening my blood and clogging my veins and arteries. Yet Tibetan populations don't have this problem. Sometime within the past 3,000 years, a gene that regulates red blood cell production spread through the population, keeping hemoglobin low, the blood flowing, and the people thriving in thin air.

Similarly, populations in Ethiopia, having settled in their high-altitude home some 5,000 years ago, also adapted. Yet others who settled up high a mere 500 years ago behave more like us lowlanders, with increased hemoglobin counts to capture as much oxygen as they can.[13] Whether or not those high-altitude genes will again sweep through the population over the next few thousand years is anyone's guess. Relatively speaking for humans, 3,000–5,000 years is a blink in evolutionary time. Each of these examples likely arose from what were once rare mutations that popped up and then spread throughout the population. Just as pesticide resistance might spread throughout weed or insect populations, rare mutations also spread through humans, except at a much slower rate because we (some more than others) are slow to mature and are far less prolific.

What are the chances that another rare mutation takes root in some human population around the world? There are now more than 7 billion of us, and with each new generation there is the potential for some 100 billion *new* mutations to arise. Most of those mutations won't do us much good, many may even be harmful. But some very small fraction—one in a billion or even a trillion—may well be advantageous.[14] With the potential for trillions of new mutations every couple of decades, we are a species rich in the raw materials necessary for evolution. And this variation, suggests genome scientist Joshua Akey, will serve us well in the long run, just as it has in the recent past. Akey suspects that traits like lactose tolerance and possibly even low-oxygen tolerance are representative of mutations that arose just as, or perhaps just before, human population growth began to accelerate. And if these kinds of traits, as Akey suggests, are representative of the types of mutations that have occurred over the past 5,000 years or so, we may be sitting on what Akey describes as a "larger reservoir of potentially adaptable alleles that can respond to environmental change."[15] In other words, according to Akey, we (as a population) may be more evolvable now than our very distant ancestors ever were.

Could some of these relatively recent yet rare mutations possibly be of benefit today? Consider the populations in Beijing, living under a life-threatening haze of air pollutants. Is it possible that a few lucky winners of the genetic lottery might be better prepared to weather the toxic storm? Perhaps a gene encoding an enzyme that quickly breaks down or sequesters toxic particulates already exists, protecting some subset of individuals from asthma or heart disease related to air pollutants. Just how quickly could a population benefit? "*In theory*," says Akey "populations can evolve really quickly. In the extreme case, think of an allele that makes an individual so fit that reproductive success is effectively guaranteed. It would take somewhere around 20 generations, depending, to sweep through the population." Twenty generations would be quick (think 20 generations of mosquitos) if we weren't talking 25 years or so per generation. And, adds Akey, those kinds of mutations—the ones that provide a really big benefit—don't tend to happen in natural populations.[16] Any reproductive advantages conferred by gene variants, even those that might provide an edge over pollutants, most likely provide only a very small fitness advantage.

And here is another reality check. We are a population of 7 billion. Even if there are some highly beneficial mutations out there, they are likely spread far and thin. A "dramatic excess of rare variation" at the population level is one thing, but whether or not selection acts upon it is really a game of probability. Even if a new gene variant protective against Beijing's toxic air exists today, what are the chances it exists in a child born in Beijing rather than a child born in Santa Fe, New Mexico, Stockholm, Sweden, or Alice Springs, Australia? Those new rare genes may exist—but unless they exist in the population under siege they won't necessarily be selected. And, restrictions on family size aside, even if such a mutation were floating around in the exposed population, if it isn't going to confer a 100 percent advantage, it will take many more than 20 generations for the new variant to sweep through the population. In other words, barring strict environmental controls, future populations in China are unlikely to be breathing easy anytime soon. Nor is it likely that our kids, grandkids, or even great-grandkids will evolve tolerance to the myriad industrial-age chemicals that we eat, drink, and breathe, even as selection works its magic on some rare genetic variant. We are cultural sprinters, but evolutionary marathoners. Unless. Unless we happen to have some *old* genes stowed away in the old genomic closet. Genes that collectively provided protection from life's hazy past: forest fires, volcanoes, or our very distant ancestor's stubborn habit of cooking over an open flame in the home cave. Perhaps a group of genes that already exist as standing genetic variation. Is it possible that this sort of standing genetic variation might, hypothetically, come in handy?

In a Blink

In 2011, Emmanuel Milot, Denis Réale, Francine M. Mayer, and colleagues investigated the possibility that traits in modern human populations could respond to selection pressure in *contemporary* time—a century or two. They discovered that age at first reproduction (when a mother births her first child) was not only influenced by natural selection but appeared to have shifted within several, rather than thousands, of generations. "Humans," they concluded "are still evolving." Their research showed that microevolutionary change—small genetic changes from one generation to the next—could be detected in a few generations, spanning just 140 years.[17] Here was evidence of human evolution happening in an astoundingly short period of time. The study population, residents of Île aux Coudres, inhabited a small island in Quebec in the St. Laurence River. This relatively self-contained population left behind 140 years of detailed genealogies beginning with the founding fathers and mothers back in 1720. "We know the founders, we know the people born there, we know the genetic relationship between these people over generations," says Milot. This kind of detail provided an opportunity for the group to tease out the genetic contribution of the decline in age at first reproduction from the cultural contribution. Age at first reproduction is a trait with strong evolutionary significance, especially in the olden days, when women who started families early in life—assuming they survived childbirth—birthed more children. Yet, acknowledges Milot, plenty of other influences come into play for traits like age at first reproduction, including environmental, social, and cultural factors. Normally it is difficult to tease these apart, but the rich genealogies of the Île aux Coudres residents provided the group with the means to "disentangle" the cultural and social from biological inheritance. In the end, Milot's group estimated that some 30 percent of the variation in the age of first reproduction was under genetic influence and, until relatively recently, subject to selection, supporting their hypothesis that evolution can indeed happen that rapidly in humans.

Any news about human evolution travels fast. "After that paper," recalls Milot, "there were all sorts of comments, pros and cons, whether we really found evolution. I just want to be clear—we support the hypothesis that this indicates evolution but this is not a definite proof." In any case, their finding—that modern human populations remain subject to natural selection—does not stand alone. The Framingham Heart Study is the longest-running health study in the United States. It includes records on the original 5,200 participants, as well as their children, and more recently their children's children. It is also a rich source of data with measurements on heritable traits in women, from height and weight to age at first reproduction and age at menopause. If we are tempted to write off Milot's population as pre-industrial, the Framingham study is *us*—our cohort of industrial-age humans. Recently, evolutionary biologist Stephen Stearns and others were able to show selection on heritable traits in the Framingham population, concluding that the evidence points to a "dynamic" rather than a "static" evolutionary status.[18] Should the selection pressures acting on these populations remain relatively constant, Stearns and others predict that female descendants will be a little shorter and stouter, will have lower total cholesterol, will have kids a bit sooner, and will have a few more years until they begin having hot flashes than they would "in the absence of evolution." This is evolution in process. Mind you, we are not talking longer thumbs or the ability for humans to suddenly find that a diet of white sugar is all they need for sustenance, but rather a few hundred grams of weight, a few millimeters in height, perhaps an extra year before menopause over the course of five or ten generations.[19] Even so, given the earlier scenario of dozens of generations *at best* before beneficial genes become fixtures in a population, how is it possible that such rapid change can occur in humans?

For continuous traits like body weight, height, and age of reproduction, while it may *seem* that evolution can be fast, selection is not acting on new or rare genes but is more likely acting on standing genetic variation. That is, selection and evolution have been working in the background all along, sort of like a computer processor. Consider all possible variability in traits rippling throughout a population, and the changing environmental, social, and cultural backdrop upon which those genes are expressed. As selection pressures change, different outcomes or phenotypes might be selected from variation existing within the population. As Milot explains, even in Île aux Coudres residents, age at first reproduction was never uniform. The *average* age may have dropped from 26 years to 22, but this represents individual ages scattered all about. It wasn't like women had to wait for a "reproduce at age 22 gene" to emerge. Unlike lactose tolerance or malaria resistance, the genetic component of age at first reproduction, as well as height, weight, and many other life-history traits, is the product of many genes, sometimes many dozens of genes. All of which exhibit a great deal of variation in a population, and this makes them more amenable to evolving rapidly—as in politics, where turning a Democrat into a Republican (or vice versa) is a rare event, but changing a *lot* of minds just a *little* bit is easier. When natural selection acts on height or weight or perhaps age at first reproduction, it may "tweak" the variants of hundreds of different genes. In effect, a small change in a lot of different genes. This is very different from most of the examples of rapid evolution discussed throughout this book. This is *not* evolution by way of a single beneficial resistance gene rising in response to an "evolve or die" situation.

So: *Are we still evolving?* The answer is a resounding yes; albeit for some traits, it is happening very slowly, while for others—continuous traits encoded by many genes—change could be quite rapid. *Is it possible we too might become less sensitive to our toxic surroundings?* As geneticist Jonathan Pritchard sums up in his article "How We Are Evolving": "The rate of change of most traits is glacially slow compared with the rate at which we change our culture and technology and, of course, our global environment. And major adaptive shifts require stable conditions across millennia."[20] In other words, when it comes to adapting to our chemical environment, we are S.O.L.

Why Not Let Nature Take Her Course?

We are changing the face of nature: adding synthetic chemicals, disrupting the climate, and dramatically altering the landscape. We have adapted our environment to suit us, but we cannot adapt quickly enough to the environment we are creating. Yet despite the resistant bedbugs, weeds, and cancers—and the increasing rates of obesity, diabetes, and infertility—it's unlikely that the human species will go the way of the dodo anytime soon. So is the guy from the arboretum right—should we simply let nature, albeit a profoundly altered nature, take its course?

I had no satisfactory answer to his question that day, and as a scientist, I still may not. But as a human being, and particularly as a mother, I worry not only about the species but about individuals. If my son gets a serious infection, I want there to be an effective antibiotic to treat him. If my daughter someday decides she wants children, I don't want her to be hampered by pesticides from supermarket produce or trace hormones and drugs in our drinking water. If our home becomes infested with bedbugs or termites, I want a lasting solution. If I or someone I love should develop cancer, and if there is no cure, then I want a drug that beats back the errant cells indefinitely. These basic matters of personal health depend on a different approach to our relationship with nature.

Like so many other species, particularly those we tend to value the most, humans are not likely to be rescued by evolution anytime soon. Instead, we are in danger of suffering not only a biological backlash from rapidly evolving species but a chemical backlash as well. Over 50 years ago, Rachel Carson observed: "As crude a weapon as the cave man's club, the chemical barrage has been hurled against the fabric of life—a fabric on the one hand delicate and destructible, on the other miraculously tough and resilient, and capable of striking back in unexpected ways."[21] In the years since Carson's dire warnings we have started to clean up our act, yet we remain a society addicted to industrial-age chemicals. Today we know far more about how these chemicals affect the environment and our own health—and the more we learn, the more reason we have for concern. It is past time to reevaluate how and when we use chemicals that are either intentionally or inadvertently toxic. Our lives and the lives of those we hold dear may well depend upon it.

PART III
Beyond Selection

CHAPTER 8

Epigenetics: Epilogue or Prologue?

The Hunger Winter was brief but devastating. Lasting from November 1944 to February1945, harsh winter conditions, combined with a Nazi blockade, caused massive starvation in occupied regions of the Netherlands. Over four million men, women, and children starved; tens of thousands died. But life went on. Women became pregnant, the next generation was born, and all of it was recorded in health registries. Decades later these registries provided invaluable insights into the interaction between environment—in this case, the nutritional environment—and genetic expression, or phenotype. For epidemiologist Lambert Lumey, molecular epidemiologist Bas Heijmans, and colleagues, the records offered a treasure trove of data on mothers, fathers, children, and grandchildren affected by the tragic conditions. From the statistics, the lasting effects of starvation began to emerge clearly: women whose mothers were malnourished while pregnant were at risk for obesity later in life; adults conceived during the famine had a higher risk of developing schizophrenia; most surprising, grandchildren born to daughters carried by mothers malnourished tended to be heavier.[1] Other than genetic mutation, which was unlikely, how could starvation influence subsequent generations well beyond those that were exposed? One answer is epigenetics: biochemical modifications that alter genetic expression but do not alter DNA sequence.

These kinds of *transgenerational* effects of environment, the nurture part of the nature–nurture equation, are turning out to be far more common than once thought and may even be relevant to evolution. Just how relevant is anyone's guess—and the question itself is enough to send some into a scientific snit, par for the course for an emerging field of study. The epigenetic story is unfolding at an increasingly rapid clip, and no doubt it will continue to do so. And so, rather than an epilogue, epigenetics may well be a prologue influencing how we think about environment and evolution.

Epigenetic Generation

Eats shoots and leaves. Eats, shoots, and leaves. Identical sets of words, two very different meanings; only the commas control how they are expressed. This is the essence of epigenetics, which describes the biochemical modification of DNA *expression* by way of chemical "marks" (small molecular tags added to DNA) or other modifications that determine whether a gene is turned on or off. Only it's a bit more complex. Some liken the genome to a musical score. The notes are represented by DNA, while the expression is governed by phrasing, how the notes are played, whether they are emphasized or not, and *how* the conductor and musicians have heard that particular score played in the past.[2]

That epigenetic modifications are at work in our cells is nothing new. Consider your own cells. Most carry identical genetic code. If you could look at the DNA sequence in your skin cells compared with that in your liver cells, it would be nearly indistinguishable. But the cell types certainly do not look or behave alike. Epigenetics explains in part how our embryonic stem cells differentiate into brain or liver or skin cells and stay that way for subsequent generations of cells. So that when you cut your finger, skin cells are replaced with skin cells, not liver or brain cells. But even into adulthood, epigenetic mechanisms are at work. As our cells age they will accrue a lifetime of environmental experiences, whether starvation, smoking, or stress, in the form of epigenetic marks. And these can be passed from one generation of cells to the next, in some cases for millions of generations—which helps explain how identical twins become increasingly different from each other as they age.[3] Some epigenetic changes contribute to cancers of, for example, the prostate, stomach, and lung; others to addictions and mental health.[4] But the marks on our somatic cells—the skin, liver, and brain—are, for the most part, ours to keep (though there is evidence that, at least in rats, *behaviors* like the grooming of pups can lead to epigenetic changes in pups—and they in turn may do the same for their offspring).[5] What reproductive biologists are finding are changes in epigenetic marks on the germ cells—the egg and sperm cells that take shape in a developing embryo, some which may be passed from one generation to the next. This is the kind of epigenetic change that can get toxicologists and evolutionary biologists alike sniping at one another.

Scientists have known for decades that epigenetic marks are carried across generations by way of germ cells. Through a complex process of biochemical choreography, epigenetic marks are wiped away and reset just prior to and then during embryonic development. This "reprogramming" (the resetting of epigenetic marks) is believed to provide the next generation with a relatively clean slate, as only marks critical for normal development are retained. But recent studies are challenging this notion and changing how we think about environmental influences on gene expression. Although there are conflicting data, a critical mass of studies is showing that some environmentally induced changes in epigenetic marks are retained through the third generation and possibly beyond in animal models. How they are transmitted isn't entirely clear. Some may insinuate themselves into the reprogramming process, while others may be retained during the "memory wipe." The findings raise profound questions about the potential for the environment to impose lasting effects on genetic expression, particularly in human populations.

Life under the (Environmental) Influence?

For children and grandchildren affected by the Hunger Winter, the timing of exposure was critical. For example, the first-born babies of *daughters* born to mothers who experienced starvation early in their development tended to be heavier.[6] Descendants of famines such as the Great Chinese Famine and the Biafra famine also showed lasting effects of starvation. Exactly how starvation influences subsequent generations isn't fully known. Whether children and grandchildren (because the developing fetus contains germ cells, a pregnant woman potentially influences the next two generations) are shaped by the *direct* or indirect effects of starvation isn't known. Lack of proper nutrition and altered stress hormones might both impact the developing fetus. Or starvation might induce epigenetic marks that can be retained through the next generation. Follow-up studies by Lumey and colleagues are finding persistent alterations in epigenetic marks of adults born to starving mothers—six decades after the event.[7] Whether or not these are transmitted to the *next* generation isn't known, but at the very least the findings demonstrate the persistence of epigenetic patterns laid down *in utero*—an epigenetic reminder of a generation's hardship. An increasing number of studies in plants and animals suggest that an array of environmental stressors, from temperature to salinity, can influence future generations—some through epigenetic mechanisms. Could toxic chemicals do the same?[8]

Working with the pesticides vinclozolin and methoxychlor, reproductive biologist Michael Skinner and colleagues have found that the chemicals can cause reduced sperm counts beyond the first and second generations into the *third* generation. Granted, the chemical concentrations associated with long-lasting effects tended to be well above the amounts found in the environment, which have caused some to question their relevance. As even Skinner admits, they'd hit the rats with a toxic hammer.[9] Even so, the results demanded attention. Subsequent studies suggest that epigenetic mechanisms may indeed be at work. And Skinner and colleagues report that dioxin and other toxic chemicals as well have been found to reach beyond the first and second generations.[10] Even as Skinner was awarded one of *Smithsonian* magazine's American Ingenuity Awards, *Science* magazine was writing about "the Epigenetics Heretic." Some say that "until there are defined mechanisms," the results will be in doubt.[11] Time and further study will surely tell.

The specter of transgenerational effects caused by toxic chemicals raises a disturbing question: *Could it happen to us?* Could the myriad synthetic and industrial chemicals flowing through our veins influence generations to come? Or are the concentrations small enough for our own defenses to kick in, maybe even kick epigenetic marks off altogether? In human studies, there is little more than association. The rise in obesity, for example, tracks with the increasing production of synthetic chemicals, but is it merely an association?[12] Is there a direct link, or is the relationship something in between? While evidence of any linkage between exposure to chemical contaminants and epigenetic effects is lacking for these effects, studies in animal models are pointing to a potential epigenetic explanation for at least some of the devastating effects in sons and daughters of women prescribed diethylstibesterol, the once-popular synthetic estrogen better known as DES.

This synthetic hormone was once mistakenly believed to help prevent miscarriage and solve a host of other hormone-related ills. Then in the 1970s, as rare vaginal cancers started showing up in daughters of DES-treated mothers, the heartbreak of DES became evident. *In utero* exposure to the synthetic estrogen caused not only cancer but a whole suite of reproductive-tract abnormalities in children. Once described as a "biological time bomb," DES may cast a larger shadow than previously known. DES *may* affect grandchildren as well.[13] While more thorough studies are under way, animal models suggest that one way the chemical may wreak havoc is through epigenetic change.[14] Studies of DES, combined with research by Skinner and others like David Crews, suggest that many different kinds of chemicals, from organochlorines, bisphenol A (BPA), and other pesticides may, like starvation, cause lasting changes in genetic expression.[15]

Even if we remain cautiously skeptical about the potential for transgenerational effects of toxic chemicals in humans, this growing body of research is bringing at least one truth to light. It is becoming increasingly clear that mothers are no longer the sole proprietors of prenatal harm; a father's prenatal and lifetime experiences can also be passed along. In other words, epigenetics (or other as-yet-unknown mechanisms) may provide fathers with an equal opportunity for transgenerational guilt.

Sperm Tales

Mothers-to-be, particularly those in their first trimester, are warned to watch their diet, drinking, stress level, and *for Pete's sake*, to stay away from any and all environmental toxicants (that goal alone is enough to raise the stress levels). It isn't all that surprising that a mother's experiences can affect her children. But, about a decade after Lumay began publishing, Marcus Pembrey added another twist: fathers matter, too, as do their experiences well before becoming fathers. Studies of descendants whose ancestors (in this case, ancestors in a remote Swedish population) had experienced variations in food availability, including periods of starvation, suggested that timing was important not only for *in utero* exposures but for adolescent exposures as well. If less food was available to the grandfather prior to puberty, his sons tended to live longer. If there was plenty, his grandsons tended toward earlier death. Again the grandparents' nutritional experiences influenced their grandchildren's health, in this case the risk of diabetes or heart disease.[16] But that wasn't all. When Pembrey and colleagues turned their attention to other early life experiences, like smoking, they found that the body mass index of young grandchildren was associated with the grandfather's smoking habit. If a grandfather smoked before age 11—that is, probably before puberty—his grandson was more likely to have a greater body mass index.[17] Others have shown that fathers who chew betel quid, an Asian stimulant, also have a lasting biological influence on their sons. However these effects are transmitted from one generation to the next—whether by way of epigenetic modifications or other yet-to-be-identified mechanisms—one thing is clear: fathers too can now be held responsible for their days of wild abandon.

Once shaken awake around the time of puberty, sperm, unlike eggs, are almost constantly dividing, making more sperm. More cell divisions generally means more opportunity for mistakes in the form of mutations. The sperm of a 70-year-old may be endowed with eight times as many mutations as the sperm of a 20-year-old; it is enough to give one pause.[18] If that's not enough, those sperm may also carry a lifetime's load of epigenetic alteration. Should those marks make it through the cell-division gauntlet, this may well explain the surprising contribution of young smokers to their grandchildren's health. But can they? And if so, how?

This is where molecular epidemiologist Rick Pilsner's research comes in. "Everyone is interested in looking at Mom's environmental health," says Pilsner, who studies the epigenetic influence of chemicals on sperm. "But no one gives consideration to the father. Does Dad also have an environmental responsibility for pregnancies and the subsequent health of the offspring?" And, adds Pilsner, what of his *lifetime* exposure? To what extent does a father's environmental history influence the next generation, and how are those effects passed down?[19] One of Pilsner's goals is to compare the epigenetic profile between a father's sperm and the cord blood of his newborn. Pilsner thinks that somewhere in the process of resetting and reprogramming there is opportunity for some epigenetic marks to escape reprogramming, passing on—for better or worse—some environmental history. The question is, asks Pilsner, "when does a 'perfect storm' of environmental insults impact the epigenome?" What makes some marks stick? While Pilsner's research, like much of the field, is a work in progress, it may well help resolve at least some of the conflict within the field.

* * *

That an experience like starvation has lasting effects is not so surprising. Natural swings in food availability, temperature, predation, and a whole host of other stressors are common for most species and were probably common in our own past (as they still are for too many). What if the offspring of a starving parent was better prepared for times of famine than a kid who grew up with plenty? It is not difficult to imagine the potential for some advantage of passing on critical environmental cues to one's offspring. But the emerging evidence that toxic chemicals can also leave their mark falls into a different category. It is difficult to conceive any good coming out of exposure to these novel environmental stressors. And—as reproductive biologist David Crews ominously reminds us—if we are talking about transgenerational effects, then we are talking about contamination across multiple generations and many different kinds of chemicals.[20] *If* chemicals leave lasting epigenetic marks, then my kids carry not only my environmental experience but reminders of their grandparents' experiences as well—a scenario that is both fascinating and disturbing. One of the most disconcerting questions may be: How long can epigenetic marks last? If marks are transferred for four, five, six generations or more—and can influence broad swaths of the population all at once—might they influence the course of evolution? In other words, how heritable are changes in epigenetic marks?

A Revolution in Evolution, or Just Another Day in the Life of DNA?

Depending on who you read, heritable epigenetics is either a missing evolutionary link, explaining variability that cannot be explained through our genes alone, or it is nothing new under the sun. Some, like Crews and endocrinologist Andrea Gore, contend that epigenetics will lead the way to a "third epoch" of evolutionary theory. This "Epigenetic Synthesis," they believe, will add environmental regulation of gene expression to the existing mix of evolutionary theory that is largely driven by the influence of gene sequence.[21] Meanwhile, others like evolutionary biologist, self-described epigenetic "curmudgeon," and blogger Jerry Coyne dismiss such revolutionary rumblings; Coyne writes: "I *know* scientific revolutions; scientific revolutions are friends of mine; and believe me, epigenetics is no scientific revolution."[22] If Crews and Gore stand at an opposite pole from Coyne, geneticist Christina Richards is more circumspect. In 2009, Richards and colleagues gathered together scientists with disparate views and experiences to hash out the role of heritable epigenetic change on the evolutionary process. Five years later, Richards says, "The jury is out. . . . There is compelling evidence suggesting that epigenetic mechanisms could be important, but there's no answer, really, about *how* important it could be."[23]

While mammalian studies indicating environmental effects stretching across to the third and fourth generations quickly become news items, plant geneticists are finding marks that last up to 20 generations.[24] But Richards warns that when it comes to genetics beyond the basics, plants and animals are very different. Many plants (but certainly not all), including many invasives like the noxious Japanese knotweed, reproduce by sending out roots, shoots, and other bits, and they don't go through the epigenetic reset as do eggs and sperm. Even when plants do engage in sex (as flowering plants are wont to do), unlike animal germ cells, plant germ cells arise from tissues that are more like our skin or liver cells—all marked up with a lifetime of environmental exposure. Plants typically throw seeds out into environment, says Richards, and they aren't as capable of changing their environment as are animals; if they don't like it, they can't just walk away. Epigenetics may provide an opportunity for differences in gene expression, ramping up a plant population's variation, which may enhance the chance of surviving different conditions, whether its seeds land at the base of the parental stock, or drift far away. But the basic differences in gamete production, epigenetic reset, and life histories between plants and animals suggest that the tenacity of epigenetic marks in plants may not translate directly or at all to animals.

That epigenetic mechanisms can so effectively and rapidly increase variation (some of which is heritable) in a population, says Richards, suggests a role for it in the selection process—but teasing out what that role is could prove incredibly difficult. Whether in a plant or an animal, the epigenome is not simply defined like the genetic code. It includes different elements and different mechanisms (e.g., chemical tags like DNA methylation, or changes in how DNA is wrapped around the spool-like histones). Recall the genome as a musical score. Just as musical expression comes and goes depending on the current musical environment, some epigenetic marks will persist while others will not. Some are attached to predictable regions of DNA and others are not. Some marks may be reversed or overridden by new marks. "Genome *function* is the holy grail," says Richards. "How does the genome manage to create these complex organisms? We just have a limited idea of how that works. Epigenetics is one part of the puzzle that holds some promise."

Epigenetics is a fast-paced and fascinating field that enhances our understanding of life and how living things interact with their environment. Whether or not epigenetics is a game-changer for evolution, it ought to be a game-changer for how we think about chemicals and chemical exposures. With a little bit of knowledge and foresight, we can prevent harm before harm becomes imprinted upon our genomes for years to come.

Notes

Chapter 1

1. For more about the history and treatment of MRSA, see: Maryn McKenna, *Superbug: The Fatal Menace of MRSA* (New York: Free Press, 2010), 1–288; A. El-Sharif and H. M. Ashour, "Community-Acquired Methicillin-Resistant *Staphylococcus aureus* (CA-MRSA) Colonization and Infection in Intravenous and Inhalational Opiate Drug Users," *Experimental Biology and Medicine* 233 (July 2008): 874–80, doi:10.3181/0711-RM-294. ("Maggie G." is a pseudonym.)
2. Centers for Disease Control and Prevention, *Antibiotic Resistance Threats in the United States 2013* (Washington, DC: US Department of Health and Human Services, 2013), 16; see also www.cdc.gov/abcs/reports-findings/survivereports/mrsa12.html.
3. Aidan Hollis and Ziana Ahmed, "Preserving Antibiotics, Rationally," *New England Journal of Medicine* 396 (December 2013): 2474–76.
4. Evan Snitkin et al., "Tracking a Hospital Outbreak of a Carbapenam-Resistant Klebsiella Pneumonia with Whole-Genome Sequencing," *Science Translational Medicine* 4 (August 2012): 148ra116, doi:10.1126/scitranslmed.3004129.
5. Dennis Pitt and Jean-Michel Aubin, "Joseph Lister: Father of Modern Surgery," *Canadian Journal of Surgery* 55 (October 2012): 1–3, doi: 10.1503/cjs007112.
6. George Sternbach, "The History of Anthrax," *Journal of Emergency Medicine* 24 (May 2003): 463–67.
7. Klaus Strebhardt and Axel Ullrich, "Paul Ehrlich's Magic Bullet Concept: 100 Years of Progress," *Nature Reviews* 8 (June 2008): 473–80.

8. Amanda Yarnell, "Salvarsan," *Chemical and Engineering News* 83 (June 2005), http://pubs.acs.org/cen/coverstory/83/8325/8325salvarsan.html.
9. Sebastian Amyes, *Magic Bullets, Lost Horizons: The Rise and Fall of Antibiotics* (London: Taylor and Francis, 2001), 8.
10. Yarnell, "Salvarsan." Interestingly, *why* Salvarsan zeros in on syphilis remains a mystery.
11. N. Svartz, "Gerhard Domagk, Presentation Ceremony Speech," in *Nobel Lectures, Physiology or Medicine 1922–1941* (Amsterdam: Elsevier Publishing Company, 1965), www.nobelprize.org/nobel_prizes/medicine/laureates/1939/press.html.
12. Alexander Fleming, "On the Antibacterial Action of Cultures of a Penicillium," *British Journal of Experimental Pathology* 10 (1929): 226–36, www.ncbi.nlm.nih.gov/pmc/articles/PMC2566493/pdf/11545337.pdf.
13. Dave Gilyeat, "Norman Heatley, the Unsung Hero Who Developed Penicillin," *BBC*, http://news.bbc.co.uk/local/oxford/hi/people_and_places/history/newsid_8828000/8828836.stm (last modified July 20, 2010).
14. Alexander Fleming, "Penicillin: Nobel Lecture," Nobelprize.org, http://nobelprize.org/nobel_prizes/medicine/laureates/1945/fleming-lecture.pdf.
15. H. Chambers and F. DeLeo, "Waves of Resistance, *Staphylococcus aureus* in the Antibiotic Era," *Nature Reviews Microbiology* 7 (September 2009): 629–41.
16. John Lesch, *The First Miracle Drug* (Oxford: Oxford University Press, 2007), 236.
17. Maureen Ogle, "Riots, Rage, and Resistance: A Brief History of How Antibiotics Arrived on the Farm," *Scientific American* (blog), September 3, 2013, http://blogs.scientificamerican.com/guest-blog/2013/09/03/riots-rage-and-resistance-a-brief-history-of-how-antibiotics-arrived-on-the-farm/.
18. E. T. Cetin and O. Ang, "Staphylococci Resistant to Methicillin ('Celbenin')," *British Medical Journal* 52 (July 7, 1962), 51–52.
19. Fred F. Barrett, MD, Read F. McGehee Jr., MD, and Maxwell Finland, MD, "Methicillin-Resistant *Staphylococcus aureus* at Boston City Hospital—Bacteriology and Epidemiologic Observations," *New England Journal of Medicine* 279 (August 1968): 441–48.

20. Centers for Disease Control and Prevention, "Antibiotic Resistance," 11.
21. Michael M. Mwangi et al., "Tracking the *In Vivo* Evolution of Multidrug Resistance in *Staphylococcus aureus* by Whole Genome Sequencing," *Proceedings of the National Academy of Science* 22 (2007): 9451–56.
22. Julian Davies, "Microbes Have the Last Word. A Drastic Reevaluation of Antimicrobial Treatment Is Needed to Overcome the Threat of Antibiotic-Resistant Bacteria," *EMBO Reports* 8 (2007): 616–21; S. Mitsuhashi et al., "On the Drug-Resistance of Enteric Bacteria," *Japan Journal of Experimental Medicine* 31 (1961): 47–52.
23. Julian Davies, "Vicious Circles: Looking Back on Resistance Plasmids," *Genetics* 139 (1995): 1465–68.
24. For a review, see: Davies, "Microbes Have the Last Word"; also see: Davies, "Vicious Circles."
25. Davies, "Vicious Circles."
26. Bacteria can acquire new genes by engaging in the processes of transformation, transduction, and conjugation. *Transformation* refers to the direct uptake of DNA from the environment (say, from a neighbor recently broken apart by penicillin). Perhaps for obvious reasons—it is not yet considered a major route for sharing resistance genes. *Transduction* is the transfer of DNA into bacteria by bacterial *viruses* or bacteriophages (phages). For example, bacterial viruses carrying resistance were recently isolated from river water and from sewage water, leading some to suggest that phages could serve as an environmental reservoir of resistance genes. For more, see: Marta Colomer-Lluch, Juan Jofre, and Maite Munieasa, "Antibiotic Resistance Genes in the Bacteriophage DNA Fraction of Environmental Samples," *PLoS ONE* 6 (March 2011): e17549, doi:10.1371/journal.pone.0017549.
27. Eugene Koonin and Yuri Wolf, "Dynamics of Bacteria and Archaea: The Emerging Dynamic View of the Prokaryotic World," *Nucleic Acids Research* 36 (2008): 6688–719.
28. Kirandeep Bhullar et al., "Antibiotic Resistance Is Prevalent in an Isolated Cave Microbiome," *PLoS ONE* 7 (April 2012): e34953, doi:10.1371/journal.pone.0034953.
29. Julian Davies, "Are Antibiotics Naturally Antibiotics?" *Journal of Industrial Microbiology and Biotechnology* 33 (March 2006): 496–99.
30. Another, more complicated role for antibiotics production may be to

enable cooperation within a bacterial population—rather than as a benefit to the individual bacterium. Just as our society is composed of warriors and nonviolent members who require protection, so, too, are populations of bacteria composed of antibiotic-producing bacteria and resistant bacteria—suggesting a kind of bacterial "social unit." For more, see: Helene Morlon, "Bacterial Cooperative Warfare," *Science* 337 (2012): 1184–85; and Otto X. Cordero et al., "Ecological Populations of Bacteria Act as Socially Cohesive Units of Antibiotic Production and Resistance," *Science* 337 (September 2012): 1228–31.

31. Julian Davies (professor emeritus in the Department of Microbiology & Immunology, University of British Columbia, Vancouver, Canada) in discussion with the author, October 2012. Note: All quoted material attributed to Julian Davies in this chapter is from this same discussion.
32. Michael Gillings and H. W. Stokes, "Are Humans Increasing Bacterial Evolvability?" *Trends in Ecology and Evolution* 27 (March 2012): 346–52.
33. Julian Davies, letter to the editor, *Globe and Mail*, April 10, 2012, http://www.theglobeandmail.com/globe-debate/april-10-letters-to-the-editor/article4098926/.
34. Karen Bush et al., "Tackling Antibiotic Resistance," *Nature Reviews Microbiology* 9 (December 2011): 894–96.
35. S. Y. Chen et al., "Health-Care-Associated Measles Outbreak in the United States after an Importation: Challenges and Economic Impact," *Journal of Infectious Disease* 203 (June 2011): 1517–25.
36. T. H. Dellit et al., "Infectious Diseases Society of America and the Society for Healthcare Epidemiology of America Guidelines for Developing an Institutional Program to Enhance Antimicrobial Stewardship," *Clinical Infectious Diseases* 44 (2007): 159–77.
37. Arjun Srinivasan, Associate Director for Healthcare and Associated Infection Prevention Programs, Division of Healthcare Quality Promotion, audio transcript for "Get Smart about Antibiotics Week 2012," United States Centers for Disease Control and Prevention, www.cdc.gov/media/dpk/2013/docs/getsmart/dpk-antibiotics-week-Arjun-Srinivasan's-audio-transcript.pdf.
38. Srinivasan, audio transcript.
39. American Academy of Microbiology, *Antibiotic Resistance: An Ecological Per-*

spective on an Old Problem (Washington, DC: American Academy of Microbiology, 2009), 11, http://academy.asm.org/images/stories/documents/antibioticresistance.pdf.

40. Dr. Arjun Srinivasan (Associate Director for Healthcare-Associated Infection Prevention Programs in the Division of Healthcare Quality Promotion at the Centers for Disease Control and Prevention) in discussion with the author, December 2013.
41. Carol Cogliani, Herman Goossens, and Christina Greko, "Restricting Antimicrobial Use in Food Animals: Lessons from Europe," *Microbe* 6 (2011): 274–79.
42. "FDA's Strategy on Antimicrobial Resistance—Questions and Answers," United States Food and Drug Association, www.fda.gov/AnimalVeterinary/GuidanceComplianceEnforcement/GuidanceforIndustry/ucm216939.htm, accessed November 2012.
43. "Protecting Employees Who Protect Our Environment," FDA Strategy Memos, PEER, www.peer.org/assets/docs/fda/10_17_12_FDA_strategy_memos.pdf, accessed November 2012.
44. For more about antibiotic use and resistance in the United States, state by state, see: The Centers for Disease Dynamics and Disease Policy, "Resistance Map" project, www.cddep.org/map.
45. H. Brotz-Oesterhelt and P. Sass, "Postgenomic Strategies in Antibacterial Drug Discovery," *Future Microbiology* 5 (October 2010): 1553–79.
46. Bush et al., "Tackling Antibiotic Resistance."
47. Srinivasan, audio transcript.
48. "Antimicrobial Resistance: Global Report on Surveillance," World Health Organization, www.who.int/drugresistance/documents/surveillancereport/en, 3, accessed May 2014.
49. B. J. *Culliton,* "Emerging Viruses, Emerging Threat," *Science 247* (January 1990): 279–80.
50. The role of indigenous bacteria in health is becoming increasingly important. Illness caused by the bacterium *Clostridium difficile,* for example, afflicts an estimated 250,000, of whom 14,000 die annually. Antibiotics are a major contributing factor by clearing out the good bacteria, leaving individuals susceptible to infection by *C. dif.* For more, see: Centers for Disease Control and Prevention, "Antibiotic Resistance." Also, recent

attempts to cure some illness by recolonizing patients with gut bacteria are proving effective. For more, see: "Quick, Inexpensive and a 90% Cure Rate," *For Medical Professionals* (blog), Mayo Clinic, www.mayoclinic.org /medical-professionals/clinical-updates/digestive-diseases/quick-inexpen sive-90-percent-cure-rate, accessed March 2014.

Chapter 2

1. Centers for Disease Control and Prevention, "Influenza Vaccination Coverage Estimates and Selected Related Results from a National Internet Panel Survey of Health Care Personnel, United States, November 2010," www .cdc.gov/flu/pdf/fluvaxview/blacknovemberhcpsurveyresults.pdf. The percentages have since increased; for more, see: Centers for Disease Control and Prevention, Morbidity and Mortality Weekly Report, "Influenza Vaccination Coverage Among Health-Care Personnel—United States, 2012–13 Influenza Season," 62 (September 27, 2013): 781–86. (Both "Annie" and "K." are pseudonyms.)
2. "First Global Estimates of 2009 H1N1 Pandemic Mortality Released by CDC-Led Collaboration," Centers for Disease Control and Prevention, www.cdc.gov/flu/spotlights/pandemic-global-estimates.htm, last updated June 25, 2012.
3. For a detailed history of virus discovery, see: Ed Rybicki and Russell Kightley, "A Short History of the Discovery of Viruses, Part 1" *ViroBlogy* (blog), February 2012, http://rybicki.wordpress.com/2012/02/06/a-short-history -of-the-discovery-of-viruses-part-1/; Ed Rybicki and Russell Kightley, "A Short History of the Discovery of Viruses, Part 2," *Viroblogy*, February 2012, http://rybicki.wordpress.com/2012/02/07/a-short-history-of-the -discovery-of-viruses-part-2/.
4. Wendell M. Stanley, "The Isolation and Properties of Crystalline Tobacco Mosaic Virus," Nobel Lecture, December 12, 1946, Nobelprize.org, www .nobelprize.org/nobel_prizes/chemistry/laureates/1946/stanley-lecture .pdf.
5. Carl Zimmer, *A Planet of Viruses* (Chicago: University of Chicago Press, 2012).
6. Luis Villarreal, "Are Viruses Alive?" *Scientific American* (December 2011): 101–5.

7. "Birth–18 Years & Catch-up Immunization Schedules," Centers for Disease Control and Prevention, www.cdc.gov/vaccines/schedules/downloads /child/0-18yrs-11x17-fold-pr.pdf, last updated January 2014.
8. Brian Deer, "How the Case Against MMR Vaccine Was Fixed," *British Journal of Medicine* 341 (2010): c6260.
9. Vincent Racaniello, "The Error-Prone Ways of RNA Synthesis," *Virology* (blog), May 10, 2009, www.virology.ws/2009/05/10/the-error-prone -ways-of-rna-synthesis/.
10. "Cumulative Number of Confirmed Human Cases of Avian Influenza A (H5N1) Reported to WHO 2003–2013," World Health Organization, March 2013, www.who.int/influenza/human_animal_interface/EN_GIP _20130312CumulativeNumberH5N1cases.pdf.
11. Ed Yong, "Mutations Behind Flu Spread Revealed," Nature News Blog, *Nature*, April 2012, http://blogs.nature.com/news/2012/04/dangerous -flu-mutations-revealed.html. For more, see: "Mutant Flu," *Nature*, www .nature.com/news/specials/mutantflu/index.html, accessed March 2014.
12. Colin Parrish (John M. Olin Professor of Virology, Cornell University, School of Veterinary Medicine, Ithaca, New York) in discussion with the author, February 20, 2013. Note: All quoted material attributed to Colin Parrish in this chapter is from this same discussion.
13. For a review, see: Colin Parrish and Yoshihiro Kawaoka, "The Origins of New Pandemic Viruses: The Acquisition of the New Host Ranges by Canine Parvoviruses and Influenza A Viruses," *Annual Review of Microbiology* 59 (2005): 553–86.
14. For a history of H1N1, see: Shanta Zimmer and Donald Burke, "Historical Perspective—Emergence of Influenza A (H1N1) Viruses," *New England Journal of Medicine* 361 (2009): 279–85.
15. For a review, see: Michael Worobey, Adam Bjork, and Joel Wertheim, "Point, Counterpoint: The Evolution of Pathogenic Viruses and Their Human Hosts," *Annual Review of Ecology, Evolution, and Systematics* 38 (2007): 521.
16. Worobey et al., "Point, Counterpoint," 532–33; for more about the potential role of viruses in myriad human diseases, see: Paul Ewald, *Plague Time* (New York: Anchor Press, 2002).
17. "Frequently Asked Questions about Multiple Vaccinations and the

Immune System," Centers for Disease Control and Prevention www.cdc.gov/vaccinesafety/vaccines/multiplevaccines.html (last updated December 2012), accessed March 2014.

18. Paul A. Offit, MD, et al., "Addressing Parents' Concerns: Do Multiple Vaccines Overwhelm or Weaken the Infant's Immune System?" *Pediatrics* 109 (January 2002): 124–29.
19. Andrew Read (Professor of Biology and Entomology and Eberly College of Science Distinguished Senior Scholar, Pennsylvania State University, State College, Pennsylvania) in discussion with the author, March 2013. Note: All quoted material attributed to Andrew Read in this chapter is from this same discussion.
20. Mark Kellet, "Rabbits in Australia: Who's the Bunny?" *Australian Heritage*, www.heritageaustralia.com.au/pdfs/Heritage%200306_Rabbits.pdf, accessed March 2014.
21. Katia Koelle (Associate Professor in Department of Biology, Duke University, Durham, North Carolina) in discussion with the author, February 2013. Note: All quoted material attributed to Katia Koelle in this chapter is from this same discussion.
22. "What You Should Know for the 2012–2013 Influenza Season," Centers for Disease Control and Prevention, www.cdc.gov/flu/about/season/flu-season-2012-2013.htm (last updated March 2013), accessed March 2014.
23. Suzanne Epstein (Associate Director for Research, Office of Cellular Tissue and Gene Therapies, Food and Drug Administration, Center for Biologics Evaluation and Research, Office of Cellular, Tissue and Gene Therapies, Division of Cellular and Gene Therapies) in e-mail discussion with the author, January 2013.
24. For more, see: Audray K. Harris et al., "Structure and Accessibility of HA Trimers on Intact 2009 H1N1 Pandemic Influenza Virus to Stem Region-Specific Neutralizing Antibodies," *Proceedings of the National Academy of Sciences* 110 (2013): 4592–97, doi:10.1073/pnas; Sabrina Richards, "Universal Flu Vaccines Charge Ahead," *Scientist* (January 2013), www.the-scientist.com/?articles.view/articleNo/33933/title/Universal-Flu-Vaccines-Charge-Ahead.
25. Nimalan Arinaminpathy et al., "Impact of Cross-Protective Vaccines on

Epidemiological and Evolutionary Dynamics of Influenza," *Proceedings of the National Academy of Sciences* 109 (2012): 3173–77, doi:10.1073/pnas.1113342109.

Chapter 3

1. Shannon Greene and Ann Reid, "Moving Targets," American Society of Microbiology, (Washington, DC, 2012), 37, http://academy.asm.org/images/stories/documents/MovingTargets.pdf.
2. For more on cancer over the ages, see: George Johnson, "Unearthing Prehistoric Tumors, and Debate," *New York Times*, December 27 2010, www.nytimes.com/2010/12/28/health/28cancer.html?_r=2&; for more about the evolutionary "mismatch" among lifestyle, diet, and health, see: Daniel Lieberman, *The Story of the Human Body: Evolution, Health, and Disease* (New York: Random House 2013); Barbara Dunn, "Solving an Age-Old Problem," *Nature* 483 (March 2012): S1–S6.
3. Jacques Robert, "Comparative Study of Tumorigenesis and Tumor Immunity in Invertebrates and Nonmammalian Vertebrates," *Developmental and Comparative Immunology* 34 (2010): 915–25.
4. Denys Wheatley, "Carcinogenesis: Is There a General Theorem?" *BioEssays* 5 (2010): 111; Carlos Sonnenschein and Ana M. Soto, "Theories of Carcinogenesis: An Emerging Perspective," *Seminars in Cancer Biology* 18 (October 2009): 372–77. For more about cancer in humans, see: Mel Greaves, *Cancer: The Evolutionary Legacy* (Oxford: Oxford University Press 2000).
5. Mark Vincent, "The Animal Within: Carcinogenesis and the Clonal Evolution of Cancer Cells Are Speciation Events *Sensu stricto*," *Evolution* 64 (April 2010): 1173–83; Robert Sanders, "Are Cancers Newly Evolved Species?" UC Berkeley News Center, July 26, 2011, http://newscenter.berkeley.edu/2011/07/26/are-cancers-newly-evolved-species/.
6. Michael R. Stratton, Peter J. Campbell, and P. Andrew Futreal, "The Cancer Genome," *Nature* 458 (2009): 719–24.
7. For an interesting discussion of cell suicide and cancer defense, see: Adam Mann, "Sponge Genome Goes Deep," *Nature* 466 (2010): 673; for more technical information, see: Tomislav Domazet-Loso and Diethard Tautz, "Phylostratigraphic Tracking of Cancer Genes Suggests a Link to the

Emergence of Multicellularity in Metazoa," *BMC Biology* 8 (2010): 66; Francesca D. Ciccarelli, "The (R)evolution of Cancer Genetics," *BMC Biology* 8 (2010): 74.

8. Douglas Hanahan and Robert A. Weinberg, "The Hallmarks of Cancer," *Cell* 100 (January 2000): 57.
9. Bernard Crespi and Kyle Summers, "Evolutionary Biology of Cancer," *Trends in Ecology & Evolution* 20 (2005): 545–52.
10. Tina Hesman Saey, "Dangerous Digs: A Cell's Surroundings May Be Instrumental to the Development of Cancer," *Science News* (October 5, 2013), www.sciencenews.org/article/dangerous-digs?mode=magazine&context=187280, accessed March 2014.
11. For more on infectious disease and cancer, see: Paul Ewald and Holly Swain, *Controlling Cancer: A Powerful Plan for Taking on the World's Most Daunting Disease* (TED Books, 2012).
12. For an interesting summary of cell life spans, see: "Life-Spans of Human Cells Defined: Most Cells Are Younger Than the Individual," *Times Higher Education*, September 2005, www.timeshighereducation.co.uk/198208.article, accessed March 2014.
13. Guy Faguet, *The War on Cancer* (Houten, the Netherlands: Springer, 2008): 69; following the war, mustard gas and other chemical weapons were outlawed by the Geneva Protocol of 1925.
14. Alfred Gilman, "The Initial Trial of Nitrogen Mustard," *American Journal of Surgery* 105 (May 1963): 574–78. Gilman noted, however, that this degree of success was not to be repeated in follow-up studies (577).
15. Ibid., 577.
16. Ibid.
17. For more about early chemotherapy treatment, see: Isaac Berenblum, *Man Against Cancer* (Baltimore: Johns Hopkins Press, 1952).
18. For more about the development of aminopterin, the folic acid analog and other anticancer treatments, see: Siddhartha Mukherjee, *The Emperor of All Maladies: A Biography of Cancer* (New York: Scribner, 2010).
19. Faguet, *The War on Cancer*, 79.
20. Ibid., 72.
21. Amir Fathi (Massachusetts General Hospital and Harvard Medical School) in discussion with the author, September 12, 2013.

22. For statistics and information on CML, see: "What Are the Key Statistics About Chronic Myeloid Leukemia?" American Cancer Society, www.cancer.org/cancer/leukemia-chronicmyeloidcml/detailedguide/leukemia-chronic-myeloid-myelogenous-key-statistics, last modified February 2014.
23. For more about treatment, see: "CML—A Short History of Treatment Since the Mid-20th Century," International Chronic Myeloid Leukemia Foundation, www.cml-foundation.org/index.php/news/169-cml-a-short-history-of-treatment-since-the-mid-20th-century, last modified June 2012.
24. R. Hehlmann et al., "Randomized Comparison of Interferon-Alpha with Busulfan and Hydroxyurea in Chronic Myelogenous Leukemia, German CML Study Group," *Blood* 84 (1994): 4064.
25. Hehlmann et al., "Randomized Comparison," 4064–77.
26. John Goldman, "Chronic Myeloid Leukemia: A Historical Perspective," *Seminars in Hematology* 47 (October 2010): 302–11.
27. Mukherjee, *Emperor*, 435.
28. Doug Jenson, "Imatinib: A Patient Perspective," in *50 Years in Hematology: Research that Revolutionized Patient Care*, American Society of Hematology, 14, www.hematology.org/About/History/50-Years/1944.aspx, accessed March 2014.
29. Ibid.; see also: E. J. Mundell, "Gleevac Continues to Beat Blood Cancer," *Washington Post*, December 6, 2006, www.washingtonpost.com/wp-dyn/content/article/2006/12/06/AR2006120601494.html.
30. Mukherjee, *Emperor*, 437.
31. Massimo Breccia and Guiliana Alimena, "Resistance to Imatinib in Chronic Myeloid Leukemia and Therapeutic Approaches to Circumvent the Problem," *Cadiovascular & Haematological Disorders—Drug Targets* 9 (2009): 21–28.
32. Michael Mathisen, Hagop Kantarjian, and Elias Jabbour, "Mutant BCR-ABL Clones in Chronic Myeloid Leukemia," *Haematologica* 96 (March 2011): 347–49.
33. Marco Gerlinger (Institute of Cancer Research, Center for Evolution and Cancer) in e-mail discussion with the author, August 2013.
34. C. Athena Aktipis, Virginia Kwan, Kathryn Johnson, Steven Neuberg, and Carlo Maley, "Overlooking Evolution: A Systematic Analysis of Can-

cer Relapse and Therapeutic Resistance Research," *PLoS ONE* 6 (November 2011): e26100, doi:10 1371/journal.pone.0026100.

35. C. Athena Aktipis (Director of Human and Social Evolution and cofounder of the Center for Evolution and Cancer at the University of California, San Francisco, and Research Professor in the Psychology Department at Arizona State University) in e-mail discussion with author, May 2014.
36. Gerlinger, discussion with author. For more about the clonal evolution of cancer, see: Peter Nowell, "The Clonal Evolution of Tumor Cell Populations," *Science* 194 (October 1976): 23–28.
37. For information on local environment and mutation, see: Bert Vogelstein et al., "Cancer Genome Landscapes," *Science* 339 (March 2013): 1546–58; for more on random mutations in tumors, see: Christoph Klein, "Random Mutations, Selected Mutations: A PIN Opens the Door to New Genetic Landscapes," *Proceedings of the National Academy of Sciences* 103 (November 2006): 18033–34.
38. Robert Gillies, Daniel Verduzco, and Robert Gatenby, "Evolutionary Dynamics of Carcinogenesis and Why Targeted Therapy Does Not Work," *Nature Reviews* 12 (July 2012): 487.
39. Gerlinger, discussion with author; in some cases, cancer stem cells may also contribute to resistance. Stem cells have yet to commit to a particular cell type (an embryonic stem cell, for example, might differentiate into a skin or muscle cell). Cancer stem cells retain the ability to differentiate into a number of different cell types within a particular tumor. Whether embryonic or cancerous, stem cells are well protected from their environment, equipped with protein pumps capable of ejecting toxic chemicals from natural metabolites to chemotherapy drugs like water from a leaky rowboat. Should a tumor be under chemical attack, stem cells may be standing by with chemically resistant recruits. The potential role of cancer stem cells is reviewed in: Darryl Shibata, "Molecular Tumor Clocks to Study the Evolution of Drug Resistance," *Molecular Pharmaceutics* 8 (2011): 2050–54; and S. Raguz and E. Yague, "Resistance to Chemotherapy: New Treatments and Novel Insights into an Old Problem," *British Journal of Cancer* 99 (2008): 387–91.
40. Mathisen et al., "Mutant BCR-ABL."
41. Jessica Cunningham, Robert Gatenby, and Joel Brown, "Evolutionary Dyna-

mics in Cancer Therapy," *Molecular Pharmaceutics* 8 (2011): 2094–100.

42. Gillies et al., "Evolutionary Dynamics," 492.
43. Ibid., 490.
44. Robert A. Gatenby, "A Change of Strategy in the War on Cancer," *Nature* 459 (May 28, 2009): 508–9.
45. Ibid.
46. Gerlinger, discussion with author.
47. For more discussion and review about resistance in agriculture, antibiotics, and chemotherapy, see: Greene and Reid, "Moving Targets," 17.
48. Francois-Xavier Mahon, "Is Going for Cure in Chronic Myeloid Leukemia Possible and Justifiable?" *American Society of Hematology Education*, Book 1 (December 2012): 122–28, doi:10.1182/asheducation-2012.1.122.

Chapter 4

1. Michael Owen (Professor of Agronomy and Extension Weed Specialist, Iowa State University, Ames, Iowa) in discussion with the author, May 2013; William Neuman and Andrew Pollack, "Farmers Cope with Round-up Resistant Weeds," *New York Times*, May 4, 2010; for details about resistant weeds, see: Ian Heap, International Survey of Herbicide Resistant Weeds, www.weedscience.com, accessed March 28, 2014.
2. 2,4-D is the common name for 2,4-Dichlorophenoxyacetic acid; atrazine is the common name for 1-Chloro-3-ethylamino-5-isopropylamino-2,4,6-triazine. Regarding recent studies into the safety of Roundup, see: T. Bohn et al., "Compositional Differences in Soybeans on the Market: Glyphosate Accumulates in Roundup Ready GM Soybeans," *Food Chemistry* 153 (2014): 207–15; Rick A. Relyea, "New Effects of Roundup on Amphibians: Predators Reduce Herbicide Mortality; Herbicides Induce Antipredator Morphology," *Ecological Applications* 22 (2012): 634–47; Rick Relyea, "Amphibians Are Not Ready for Roundup," *Wildlife Ecotoxicology* (*Emerging Topics in Ecotoxicology* series) 3 (201): 267–300.
3. H. G. Baker, "Characteristics and Modes of Origin of Weeds," in H. G. Baker and G. L. Stebbins, *The Genetics of Colonizing Species, Proceedings of the First International Union of Biological Sciences Symposia on General Biology, Asilomar, California* (New York: Academic Press, 1965): 147–72.
4. Edward O. Wilson, *The Diversity of Life* (New York: W. W. Norton, 1993), 300.

5. Fred Gould, "The Evolutionary Potential of Crop Pests," *American Scientist* 79 (November-December 1991): 497.
6. EPSPS is 5-enolpyruvylshikimate-3-phosphate synthase.
7. One study in particular, conducted by Gilles-Eric Seralini and colleagues, which reported liver and kidney toxicity as well as tumor development following exposure to Roundup Ready corn cultivated with and without Roundup, has created much controversy. The study was eventually retracted. For letters regarding the retraction, see: Gilles-Eric Seralini et al., "RETRACTED: Long-Term Toxicity of a Roundup Herbicide and a Roundup-Tolerant Genetically Modified Maize," *Food and Chemical Toxicology* 50 (November 2012): 4221–31. For more about the controversy, see: "Seralini Affair," Wikipedia, http://en.wikipedia.org/wiki/S%C3%A9ralini_affair, accessed March 31, 2014. Another study hypothesized that Roundup may be related to a high incidence of kidney toxicity; for more, see: "Glyphosate, Hard Water, and Nephrotoxic Metals: Are They the Culprits Behind the Epidemic of Chronic Kidney Disease of Unknown Etiology in Sri Lanka?" *International Journal of Environmental Health and Public Health* 11 (2014): 2125–47.
8. For a history of no-till farming, see: http://research.wsu.edu/resources/files/no-till.pdf. No-till farming was enabled a decade earlier by "chemical seed-bed" preparation and a combination of herbicides including paraquat, 2,4-D, and atrazine.
9. For a detailed history, see: Daniel Charles, *Lords of the Harvest* (Cambridge, MA: Perseus, 2001).
10. Paul Berg, "Asilomar and Recombinant DNA," Nobelprize.org, Nobel Media AB, 2013, www.nobelprize.org/nobel_prizes/chemistry/laureates/1980/berg-article.html, accessed March 28, 2014.
11. Charles, *Lords of the Harvest*, 126–48.
12. Ibid., 42.
13. Ibid., 68.
14. Diane Re and Stephen Padgette, "Petition for Determination of Nonregulated Status: Soybeans with a Roundup Ready™ Gene," USDA Plant and Animal Inspection Services Documents www.aphis.usda.gov/brs/aphisdocs/93_25801p.pdf, accessed March 28, 2014; A. McHughen and S. Smyth, "US Regulatory System for Genetically Modified [Genetically

Modified Organism (GMO), rDNA or Transgenic] Crop Cultivars," *Plant Biotechnology Journal* 6 (2008): 2–12.

15. Ibid., 3.
16. Jonathan Gressel, "Evolving Understanding of the Evolution of Herbicide Resistance," *Pest Management Science* 65 (2009): 1171.
17. Ian Heap, "International Survey of Resistant Weeds," www.weedscience.com/summary/home.aspx, accessed March 31, 2014.
18. Gressel, "Evolving Understanding," 1166.
19. Robert Gordon Harvey, letter to Diane Re, February 1993, www.aphis.usda.gov/brs/aphisdocs/93_25801p.pdf, accessed March 31, 2014.
20. Laura Bradshaw et al., "Perspectives on Glyphosate," *Weed Technology* 11 (1997): 189.
21. Michael Owen, letter to Diane Re, February 8, 1993, www.aphis.usda.gov/brs/aphisdocs/93_25801p.pdf, accessed March 31, 2014.
22. Charles Benbrook, "Impacts of Genetically Engineered Crops on Pesticide Use in the U.S.—The First Sixteen Years," *Environmental Sciences Europe* 24 (2012): 4, doi:10.1186/2190-4715-24-24.
23. For sales figures, see: Unites States Environmental Protection Agency, "2006–2007 Pesticide Market Estimates," www.epa.gov/opp00001/pestsales/07pestsales/table_of_contents2007.htm, accessed March 31, 2014; "Global Glyphosate Market to Reach 1.35 Million Metric Tons by 2017," PRWeb.com, www.prweb.com/releases/glyphosate_agrochemical/technical_glyphosate/prweb8857231.htm, accessed March 31, 2014.
24. "Roundup Ready System, Overview," Monsanto, www.monsanto.com/weedmanagement/Pages/roundup-ready-system.aspx, accessed March 31, 2014.
25. Charles Arntzen, Welcome Message, in *Proceedings of the National Summit on Strategies to Manage Herbicide-Resistant Weeds* (Washington, DC: National Academies Press, 2012), 2.
26. Vijay Nandula et al., "Glyphosate-Resistant Weeds: Current Status and Future Outlook," *Outlooks on Pest Management* (August 2005): 183–87, doi:10.1564/16aug11.
27. Council for Agricultural Science and Technology (CAST), "Herbicide-Resistant Weeds Threatens Soil Conservation Gains: Finding a Balance for Soil and Farm Sustainability," Issue Paper 49 (February 2012), 13.

28. Owen, discussion with author.
29. "We prefer to educate rather than regulate," says Rick Cole, Monsanto's technology application director, as quoted as by Harry Cline in "Too Many Stacked Crop Trait Genes?" Western Farm Press, April 2010, http://westernfarmpress.com/management/too-many-stacked-crop-trait-genes, accessed May 2014.
30. Greg Jaffe (Biotechnology Project Director for the Center for Science in the Public Interest), in discussion with the author, June 2013.
31. Ibid.
32. Ibid.
33. David Mortensen et al., "Navigating a Critical Juncture for Sustainable Weed Management," *BioScience* 62 (2012): 82, doi:10.1525/bio.2012.62.1.12.
34. For more, see: Mortensen et al., "Navigating a Critical Juncture," 75–84; see also: Heap, "International Survey."
35. Mortensen et al., "Navigating a Critical Juncture," 75.
36. Department of Agriculture, Animal Plant Health Inspection Service, Notice [Docket] No. APHIS-2013-0042, *Federal Register* 78 (Thursday, March 16, 2013): 28798–800, www.gpo.gov/fdsys/pkg/FR-2013-05-16/html/2013-11579.htm, accessed March 31, 2014.
37. United States Environmental Protection Agency, "EPA Seeks Comment on Proposed Decision to Register Enlist," www.epa.gov/pesticides/factsheets/2-4-d-glyphosate.html#weeds, accessed June 6, 2014.

Chapter 5

1. Michael Potter, Jim Fredericks, and Missy Henriksen, "2013 Bugs Without Borders Survey," Pestworld.org, www.pestworld.org/news-and-views/pest-articles/articles/2013-bugs-without-borders-survey-executive-summary, accessed April 1, 2014.
2. For news stories on bedbug infestations, see: Joe Pompeo, "Where's Roscoe? Bedbugs Infest Wall Street," *Capital New York*, www.capitalnewyork.com/article/media/2013/01/7174682/wheres-roscoe-bedbugs-infest-wall-street-journal-headquarters, accessed April 1, 2014; and: Joseph Berger, "Bedbugs and Worried Workers at Anything But Fleabag Hotel," *New York Times*, January 25, 2012, www.nytimes.com/2012/01/26/nyregion

/bedbug-at-ritz-carlton-alarms-hotel-worker.html?_r=2&partner=rss&emc=rss&.

3. Alvaro Romero (Assistant Professor of Urban Entomology, New Mexico State University, Las Cruces, New Mexico) in discussion with the author, August 2013.
4. "Bedbug Registry," Bedbugregistry.com, accessed April 1, 2014.
5. T. G. E. Davies, L. M. Field, and M. S. Williamson, "The Re-emergence of the Bed Bug as Nuisance Pest: Implications of Resistance to the Pyrethroid Insecticides," *Medical and Veterinary Entomology* 26 (2012): 242. Having maintained bedbug colonies on his own blood for decades, Harold Harlan has provided an opportunity for a new generation of researchers to work with bedbugs. Harlan's bugs are now sought after by researchers seeking non-resistant populations; for more, see: "Let the Bed Bugs Bite," *Science News*, www.sciencenews.org/article/let-bedbugs-bite, accessed April 1, 2014.
6. Michael Potter, "The History of Bed Bug Management—With Lessons from the Past," *American Entomologist* (Spring 2011): 14–25.
7. Coby Schal, "Importance of Basic and Translational Approaches to Bed Bug Management," Bed Bug 2011 Summit presentation, www.epa.gov/pesticides/ppdc/bedbug-summit/2011/21-scha-bb.pdf, accessed April 1, 2014.
8. Potter, "The History," 16.
9. Ibid., 18.
10. Paul Müller, "Dichloro-diphenyl-trichloroethane and New Insecticides," Nobel Prize Lecture, www.nobelprize.org/nobel_prizes/medicine/laureates/1948/muller-lecture.pdf, accessed April 1, 2014.
11. Thomas Dunlap, *DDT: Scientists, Citizens, and Public Policy* (Princeton, NJ: Princeton University Press, 1981), 37.
12. "DDT Regulatory History: A Brief Survey (to 1975)," from *DDT, A Review of Scientific and Economic Aspects of the Decision to Ban Its Use as a Pesticide*, report prepared for the Committee on Appropriations of the US House of Representatives by EPA, July 1975, EPA-540/1-75-022, www2.epa.gov/aboutepa/ddt-regulatory-history-brief-survey-1975, accessed April 1, 2014.
13. Dini Miller and Stephen Kells, "Bed Bug Action Plan for Home Health Care and Social Workers," Virginia Department of Agriculture and Con-

sumer Services, www.vdacs.virginia.gov/pesticides/pdffiles/bb-healthcare1.pdf, accessed April 1, 2014.

14. Potter, "The History," 22.
15. James Busvine, "Insecticide Resistance in Bed Bugs," *Bulletin of the World Health Organization* 19 (1958): 1041–52.
16. Gary Thompson, "MSU Resistance Database," 46th meeting of IRAC [Insecticide Resistance Action Committee] International, www.irac-online.org/documents/resistance-database-team-update-2011/?ext=pdf, accessed April 1, 2014.
17. Some formulations include a metabolic inhibitor, extending the chemical's efficacy.
18. Fang Zhu, "Evolved Unique Adaptive Strategy to Resist Pyrethroid Insecticides," *Scientific Reports* 3 (March 2013), doi:10.1038/srep01456; Alvaro Romero et al., "Insecticide Resistance in the Bed Bug: A Factor in the Pest's Sudden Resurgence?" *Journal of Medical Entomology* 44 (2007): 175–78.
19. Fang Zhu et al., "Bed Bugs Evolved Unique Adaptive Strategy to Resist Pyrethroid Insecticides," *Scientific Reports* 3 (March 2013): 1456, doi: 10.1038/srep01456, accessed June 3, 2014.
20. Megan Szyndler et al., "Entrapment of Bed Bugs by Leaf Trichomes Inspires Microfabrication of Biomimetic Surfaces," *Journal of the Royal Society Interface* 10 (April 2013): 1–9, doi:10.1098/rsif.2013.0174, accessed April 1, 2014; Felicity Barringer, "How a Leafy Folk Remedy Stopped Bedbugs in Their Tracks," *New York Times*, April 9, 2013, A16.

Chapter 6

1. Isaac Wirgin (Associate Professor, Department Environmental Medicine, New York University, Langone Medical Center, Tuxedo, New York) in discussion with the author, November 2013.
2. John Bickham and Michael Smolen, "Somatic and Heritable Effects of Environmental Genotoxins and the Emergence of Evolutionary Toxicology," *Environmental Health Perspectives* 102 (Supplement 12, December 2012): 27.
3. Claude Boyd (Professor of Limnology and Water Quality in Aquacul-

ture, Auburn University, Auburn, Alabama) in discussion with the author, November 2013.

4. Rachel Carson, *Silent Spring* (Boston: Houghton Mifflin, reprint: 1994), 271.
5. Ibid., 272.
6. Claude Boyd, S. Bradleigh Vinson, and Denzel Ferguson, "Possible DDT Resistance in Two Species of Frog," *COPIA* 2 (1963): 428–29.
7. C. Mary Boyle, "Case of Apparent Resistance of *Rattus norvegicus* Berkenhout to Anticoagulant Poisons," *Nature* 188 (November 1960): 517; G. W. Ozburn and F. O. Morrison, "Development of a DDT-Tolerant Strain of Laboratory Mice," *Nature* 196 (December 1962): 1009–10.
8. Claude Boyd and Denzel Ferguson, "Susceptibility and Resistance of Mosquito Fish to Several Insecticides," *Journal of Economic Entomology* 57 (1964): 430–31; Boyd et al., "Possible DDT Resistance." Recent studies in wood frogs suggest evolution of resistance in that species; for more, see: Rickey Cothran, Jenise Brown, and Rick Relyea, "Proximity to Agriculture Is Correlated with Pesticide Tolerance: Evidence for the Evolution of Amphibian Resistance to Modern Pesticides," *Evolutionary Applications* 6 (2013): 832–41.
9. Ozburn and Morrison, "Development of a DDT-Tolerant Strain of Mice."
10. Andrew Whitehead et al., "Common Mechanism Underlies Repeated Evolution of Extreme Pollution Tolerance," *Proceedings of the Royal Society B* 279 (2012): 427–33.
11. Peter Grant and B. Rosemary Grant, "Predicting Microevolutionary Response to Directional Selection on Heritable Variation," *Evolution* 49 (1995): 241–51; Peter Grant and B. Rosemary Grant, "Unpredictable Evolution in a 30-Year Study of Darwin's Finches," *Science* 296 (2002): 707–11.
12. Ibid.
13. Steven Brady, "Road to Evolution? Local Adaptation to Road Adjacency in an Amphibian *(Ambystoma maculatum)*," *Scientific Reports* 2 (2012), doi: 10.1038/srep00235.
14. J. R. Mullaney, D. L. Lorenz, A. D. Arntson, "Chloride in Groundwater and Surface Water in Areas Underlain by the Glacial Aquifer System, Northern United States, U.S. Geological Survey Investigations Report 2009-5086.

15. Steven Brady (Postdoctoral Associate, Dartmouth College) in discussion with the author, May 2012.
16. Brady, discussion with author.
17. Paul Klerks (Department of Biology, University of Louisiana) in discussion with the author, December 2013.
18. See: Ewen Callaway, "Inbred Royals Show Traces of Natural Selection," *Nature* News (April 19, 2012), doi:10.1038/nature.2013.12837.
19. Klerks, discussion with author.
20. Sébastien Bélanger-Deschênes et al., "Evolutionary Change Driven by Metal Exposure as Revealed by Coding SNP Genome Scan in Wild Yellow Perch (*Perca flavescens*)," *Ecotoxicology* (May 2013), doi:10.1007/s10646-013-1083-8.
21. Lingtian Xie and Paul Klerks, "Fitness Cost of Resistance to Cadmium in the Least Killifish (*Heterandria formosa*)," *Environmental Toxicology and Chemistry* 23 (2004): 1499–1503.
22. T. M. Uren Webster et al., "Global Transcriptome Profiling Reveals Molecular Mechanisms of Metal Tolerance in a Chronically Exposed Wild Population of Brown Trout," *Environmental Science and Technology* 47 (2013): 8869–77.
23. Peter Rosato and Denzel Ferguson, "The Toxicity of Endrin-Resistant Mosquitofish to Eleven Species of Vertebrates," *BioScience* 18 (1968): 783–84.
24. "PCB Contamination of the Hudson River Ecosystem Compilation of Contamination Data through 2008," *Hudson River Natural Resource Damage Assessment*, report of Hudson River Natural Resource Trustees (New York, January 2013), www.fws.gov/contaminants/restorationplans/hudsonriver/docs/Hudson%20River%20Status%20Report%20Update%20January%202013.pdf, accessed April 1, 2014.
25. Ibid.
26. Judith Weis (Professor Emerita, Department of Biological Sciences, Rutgers College of Arts and Sciences, Newark, New Jersey) in discussion with the author, December 2014. Weis finds increased tolerance in killifish embryos compared with cleaner sites, but she has not confirmed that these traits are heritable, and so cannot confirm the likelihood of evolutionary change.

27. Anthony Barnosky et al., "Has the Earth's Sixth Mass Extinction Already Arrived?" *Nature* 471 (March 2011): 56.
28. Ibid., 51–57.

Chapter 7

1. Bill McKibben, *Eaarth: Making a Life on a Tough New Planet* (New York: Henry Holt and Company, 2010.)
2. Barbara King, "Attenborough's Muddled Thinking Can't Stop Human Evolution," *NPR Cosmos and Culture*, September 12, 2013, www.npr.org/blogs/13.7/2013/09/12/221719438/attenboroughs-muddled-thinking-cant-stop-human-evolution; Ian Ricard, "Sir David Attenborough Is Wrong—Humans Are Still Evolving," *Guardian* (blog), September 10, 2013, www.theguardian.com/commentisfree/2013/sep/10/david-attenborough-humans-still-evolving; Madhusudan Katti, "Sorry Attenborough, Humans Still Evolve by Natural Selection," *The Conversation* (blog), September 18, 2013, https://theconversation.com/sorry-attenborough-humans-still-evolve-by-natural-selection-18124; "They Said It," *Science*, 341 (September 2013): 1325.
3. For a review, see: Russell Powell, "The Future of Human Evolution," *British Journal for the Philosophy of Science* 63 (2012): 145–75.
4. Steven Jones, "Enlightenment Lecture—Is Human Evolution Over?" video, September 25, 2009, www.youtube.com/watch?v=XE_Oy1eRyVg, accessed April 3, 2014; Ian Tattersall, interview on *DNATube*, www.dnatube.com/video/682/Human-evolution-interview-with-Ian-Tattersall, accessed April 3, 2014.
5. "How *Homo sapiens* Became 'Masters of the Planet,'" Ira Flatow interview with Ian Tattersall, National Public Radio, April 6, 2012, www.npr.org/2012/04/06/150123937/how-homo-sapiens-became-masters-of-the-planet.
6. Jones, "Enlightenment Lecture."
7. Denis Réale, "Human Change We Can Believe In," *Project Syndicate* (blog), August 24, 2012, www.project-syndicate.org/commentary/human-change-we-can-believe-in-by-denis-reale#9Jcbtp7FCmO2qZif.99.
8. Michael Balter, "Are Humans Still Evolving?" News Focus, *Science* 309 (July 2005): 234.

9. Emmanuel Milot (University of Sherbrooke, Department of Medicine, Quebec, Canada) in discussion with the author, January 2014.
10. Emmanuel Milot and Fanie Pelletier, "Human Evolution: New Playgrounds for Natural Selection," *Current Biology* 23 (2013): R448.
11. "City Living Helped Humans Evolve Immunity to Tuberculosis and Leprosy, New Research Suggests," *Sciencedaily* (blog), September 24, 2010, www.sciencedaily.com/releases/2010/09/100923104140.htm.
12. James Owen, "Stone Age Milk Use Began 2,000 Years Earlier," *National Geographic News* (blog), August 6, 2008, http://news.nationalgeographic.com/news/2008/08/080806-prehistoric-dairy.html.
13. "Researchers Solve Questions about Ethiopians' High-Altitude Adaptations," *ScienceDaily* (blog), January 12 2012, www.sciencedaily.com/releases/2012/01/120120184530.htm; "Ethiopians and Tibetans Thrive in Thin Air Using Similar Physiology, but Different Genes," *Eureka Alert*, National Evolutionary Synthesis Center news release, December 6, 2012, www.eurekalert.org/pub_releases/2012-12/nesc-eat120612.php.
14. Wenquing Fu and Joshua Akey, "Selection and Adaptation in the Human Genome," *Annual Review of Genomics and Human Genetics* 14 (August 2013): 467–89.
15. Joshua Akey (Associate Professor of Genome Science, University of Washington, Seattle, Washington) in discussion with the author, January 2014. Note: All quoted material attributed to Joshua Akey in this chapter is from this same discussion.
16. Akey, discussion with author.
17. Milot, discussion with author; Emmanuel Milot et al., "Evidence for Evolution in Response to Natural Selection in a Contemporary Human Population," *Proceedings of the National Academy of Sciences* 108 (October 2011): 17040–45.
18. Steven Stearns et al., "Measuring Selection in Contemporary Human Populations," *Nature Reviews Genetics* 11 (September 2010): 611.
19. Sean Byers et al., "Natural Selection in a Contemporary Human Population," *Proceedings of the National Academy of Sciences* 107, Supplement 1 (January 2010): 1787–92.
20. Jonathan Pritchard, "How We Are Evolving," *Scientific American* (October 2010): 47.

21. Rachel Carson, *Silent Spring* (Boston: Houghton Mifflin, reprint edition, 1994), 296.

Chapter 8

1. L. H. Lumey et al., "Cohort Profile: The Dutch Hunger Winter Families Study," *International Journal of Epidimiology* 36 (2007): 1196–1204; Nessa Carey, *The Epigenetics Revolution* (New York: Columbia University Press, 2012): 102.
2. Eva Jablonka and Marion Lamb, "Evolution in Four Dimensions" (Cambridge, MA: MIT Press, 2005): 110.
3. Richard Stein, "Epigenetics and Environmental Exposures," *Journal of Epidemiology and Community Health* 66 (2012): 8–13.
4. Eric Nestler, "Epigenetic Inheritance: Fact or Fiction?" *The Dana Foundation* (blog), April 2013, www.dana.org/Publications/ReportOnProgress/Epigenetic_Inheritance_Fact_or_Fiction/.
5. Reviewed in Catherine Aiken and Susan Ozanne, "Transgenerational Developmental Programming," *Human Reproduction Update* 20 (2014): 63–75.
6. Carey, *The Epigenetics Revolution*, 102.
7. Bastiaan Heijmans et al., "Persistent Epigenetic Differences Associated with Prenatal Exposure to Famine in Humans," *Proceedings of the National Academy of Sciences* 105 (November 2008): 17046–49.
8. Aiken and Ozanne, "Transgenerational Developmental Programming."
9. Jeneen Interlandi, "Code Breaker," *Smithsonian* (December 2013): 78–82.
10. Mohan Manikkam et al., "Dioxin (TCDD) Induced Epigenetic Transgenerational Inheritance of Adult Onset Disease and Sperm Epimutations," *PLoS ONE* 7 (September 2012): e46249.
11. Quoted in Jocelyn Kaiser, "The Epigenetics Heretic," *Science* (January 2014): 362.
12. P. F. Bialle-Hamilton, "Chemical Toxins: A Hypothesis to Explain the Global Obesity Epidemic," *Journal of Alternative and Complimentary Medicine* 8 (2002): 185–92.
13. John McLachlan, "Commentary: Prenatal Exposure to Diethylstilbesterol (DES): A Continuing Story," *International Journal of Epidemiology* 35 (2006): 868.

Island Press | Board of Directors